A Technical Handbook
on
Bituminized Jute Paving Fabric (BJPF) – a partial substitute and reinforcement of bitumen mastic

A Technical Handbook
on
Bituminized Jute Paving Fabric (BJPF) – a partial substitute and reinforcement of bitumen mastic

Swapan Kumar Ghosh

WOODHEAD PUBLISHING INDIA PVT LTD

New Delhi, India

Published by Woodhead Publishing India Pvt. Ltd.
Woodhead Publishing India Pvt. Ltd., 303, Vardaan House, 7/28, Ansari Road,
Daryaganj, New Delhi - 110002, India
www.woodheadpublishingindia.com

First published 2016, Woodhead Publishing India Pvt. Ltd.
© Woodhead Publishing India Pvt. Ltd., 2016

Reprint 2020

Woodhead Publishing India Pvt. Ltd. ISBN: 978-9-38505-919-3
Woodhead Publishing Ltd. e-ISBN: 978-9-38505-965-0

Typeset by Third EyeQ Technologies Pvt Ltd, New Delhi
Printed and bound by Replika Press Pvt. Ltd.

Contents

Acknowledgement

The author expresses his sincere gratitude to Jute Technology Mission (Mini Mission – IV, Scheme- 7.1), sponsored by National Jute Board (NJB), Ministry of Textiles, Government of India for providing the platform to the authors to fabricate the paper. The author is indebted to the then Hon'ble Secretary Shri Arijit Banerjee I.F.S., for his valued support at various stages of the progress of this research work. The author is very much thankful to Shri Tapobrata Sanyal, Chief Consultant, National Jute Board (NJB), Ministry of Textiles, Government of India for his consistent cooperation and wise guidance throughout in successfully shaping this book. Thanks are also due to Shri D.C. Baheti, Executive Director, M/s. Gloster Limited, 21 Strand Road, Kolkata-700 001 and M/s. Gladstone Lyall Industrial Co-operative Society Ltd., Kolkata and Indian Jute Mills Association (IJMA), India. The author is also thankful to the Director, Central Road Research Institute (CRRI), New Delhi, India.

The author express his heartiest thanks to the then Hon'ble Vice Chancellor University of Calcutta and Hon'ble Mayor, Kolkata, India for their kind permissions to carry out this research work alongwith its commercial field trial respectively.

Finally, the author extends his thanks and affection to Shri Murari Mohan Mondal and Shri Rajib Bhattacharyya, Senior Research Fellows and Teaching Associates of Department of Jute and Fibre Technology, Institute of Jute Technology, University of Calcutta, West Bengal, India, along with Shri Sambhu Nath Dey, Laboratory Attendant of the Department for their consistent support and contribution in accolading this research work into the shape of a book.

Preface

Technology and progress of civilization, economic development and environment, the imperceptible world of economics and infrastructure development do not go hand in hand. There is a dire need for the biological systems to remain diverse and productive indefinitely, cajoled with wise application of technology. The leap into the future could well be our farewell if we fail to realize our moral responsibility towards the rich natural resources we are gifted with and to use technology in line.

In the last quarter of the twentieth century, a new class of materials, called, Geosynthetics, emerged which led to significant revolution in the design of geotechnical and geoenvironmental systems.

I got engaged in extensive research and developmental work related to the field of Geosynthetics and Geotextile Technology since 1997 when Geotextiles both in man-made and natural fibres started proving effective in improving geotechnical characteristics of soil and found extensive uses for various technical end-uses like erosion control, management of slopes, strengthening of roads, stabilization of embankments, protection of river-banks, consolidation of soft soil, etc.

Over the years intensive studies have shown Geotextiles to prove amongst the most versatile and techno-economically viable ground modification materials playing a significant role in modern pavement design and maintenance techniques. With the growing environmental concern across the globe, technologists, researchers and engineers have inclined towards the natural geotextile where Jute Geotextile (JGT) is one of the most promising and potential candidates. But, JGT has been restricted mainly as underlay in road construction. Hence, I was feeling an urge that there is an urgent need to design and develop a precise innovative fabric as overlay on existing pavements and other emerging civil works to stay technically and economically competitive in the global market.

Geotextiles extend the service life of roads, increase their load carrying capacity and reduce rutting and other distresses. The effectiveness of geotextiles in stabilization and separation roles with flexible pavements has been extensively researched. Asphalt concrete pavements overlays can benefit from the use of paving fabric interlayer. To make effective use as an overlay fabric on existing pavements, I researched and found that paving fabric has to be water-proofed and abrasion resistant. It has been reported from the several case studies and experiments that non-woven jute geotextile is an extremely good receptor of hot bitumen, besides having thermal compatibility with bitumen in the range of 190 °C. Woven Jute Geotextile ensures durability against abrasion and shear. Hence, the idea of a combination of woven and non-woven jute fabric was an eye-opener to me which when smeared with suitable type and grade of bitumen can be used as an overlay fabric for strengthening of the pavements as well as partial substitute of commonly used bitumen mastic in road construction. To achieve this objective, I carried out exhaustive research and development activities over the last five years as to how best the material could be put use for various geotechnical applications followed by numerous field applications. The research work turned out to be productive and effective with the production of Grey Jute Paving Fabric (GJPF) samples, followed by its bituminization with suitable type and grade of bitumen, and several field trials of pilot and commercial levels to assess the performance and techno-economic viability of the developed bituminized jute paving fabric to establish its efficacy as a potential innovative civil engineering material.

Recognizing the vast potentiality of this developed Bituminized Jute Paving Fabric (BJPF) as overlay on existing pavements to reinforce and partially substitute the bitumen mastic a strong desire of sharing my findings stimulated me to write this comprehensive technical handbook. I firmly believe that this book will at least partially fulfil the requirements of the interested engineering students and practising engineers and may prompt them to delve deeper into the subject to explore new avenues for its use and to refine the existing design methodologies.

I would feel amply rewarded if the technical handbook can be a bridge between the practising civil engineers and jute technologists generating interest amongst them.

Swapan Kumar Ghosh

Professor and Head

Department of Jute and Fibre Technology.

Introducing geotextile

1.1 Introduction

In the last quarter of the last century, a new class of materials, called geosynthetics, emerged worldwide which led to significant innovation in the design of geotechnical and geoenvironmental systems. The proliferation of geosynthetics in reinforced soil technology has revolutionized the way in which the practicing civil and geotechnical engineers now think of walls and embankments. In view of the importance being given to rapid infrastructure development, be it roads and bridges, ports and waterways, or municipal and hazardous waste landfills, geosynthetics have found their way into civil engineering constructions in India. Today it has become evident that the benefits include economy, faster construction while simultaneously dealing with difficult soil conditions. With the large variety of geosynthetics now available, an astonishingly wide spectrum of application of 'Geotextile'[1] took up the task of providing a lucid solution to address the several soil related problems. Geotextiles, a class of geosynthetics, and belonging to the category of technical textiles (Geotech) is defined by ASTM D4439[2] as 'A permeable geosynthetic comprised solely of textiles' that are used in or on soil, rock, earth or any other geotechnical related material as an integral part of civil engineering project, structure or system to improve the soil's engineering performance. Control of soil erosion and earth slips, prevention of soil migration from under the base of a structure, dissipation of water from soil-body, separating soil layer from overlying courses, strengthening of roads, stabilization of embankments, protection of river banks, consolidation of soft soil are the critical functions that geotextiles perform to improve the engineering performance of soil.[3] In performing these functions geotextiles have proven to be among the most versatile and cost effective ground modification materials and their use has extended rapidly into nearly all areas of civil, geotechnical, environmental, coastal and hydraulic engineering. The role of geotextile in addressing the host of soil related problems is that of a change agent.

1.2   Historical background

Geotextiles have been used in several forms for thousands of years which include geomembranes, geogrids, geonets, geomeshes, geomatts and geocomposites.[4] Interestingly, the concept of laying man-made geotextiles on soil was not a new invention in the strict sense. Historical records reveal that compacted soil reinforced with river reeds was in use in the Iranian Plateau way back in the fifth millennium BC.[5] These reed-grids had long been in use for coastal and river bank erosions and can be aptly called precursors to present day geotextiles. Even in the late nineteenth century, 'faggots' or 'fascines' made of well-ordered bundles of willow, alder or brushwood were used in construction of levees (Wheeler 1893; Strombusijsing 1854). In the Mississippi River, fascine mattresses were used for control of bed erosion. The first use of a textile fabric structure for geotechnical engineering was in 1926 for road construction. It is worthy to mention here, that geotextiles made of jute were tried in Dundee, Scotland (Kingsway Road) in 1920 and in Calcutta, India (Strand Road) in 1934 long before the conceptualization of man-made geotextiles.[?]In 1930s woven jute fabric was used for sub-grade support in construction of highway in Aberdeen. The maiden use of a woven synthetic fabric for erosion control was in 1950s in Florida by Barrett. In 1960s geotextiles were extensively used for erosion control both in Europe as well as USA. Also the first use of non-woven fabric in civil engineering was for asphalt overlay in USA in the year 1966.[6] In 1962, M/s Nelton Ltd., UK, used synthetic nets for the first time in a civil engineering project. Similarly, reinforcement work of soft ground in Japan was conducted under the responsibility of Professor I. Yamanauchi. This successful trial was followed by many applications, including embankment reinforcement for the Japanese National Railway and inspired the development of geogrids. In 1969, Giroud had used non-woven fabrics as a filter in the upstream face of an earthen dam.[7] In 1971, Wager initiated use of woven fabrics as reinforcement for embankments constructed on very soft foundations. More recently in the nineteenth century, George Stephenson used fibrous materials, including waste cotton bales, to provide a water permeable and flexible foundation for the world's first passenger railway in UK. With such humble beginnings geotextiles/ geosynthetics or related products are being increasingly used all over the world in the field of civil engineering construction. It can be considered that the 'first generation' geotextiles were textiles that were being manufactured for other purposes (such as carpet or industrial sackings) but which were diverted and used for geotechnical purposes.[8] The 'second generation' geotextiles were produced by choosing specific textiles suitable for geotechnical purposes in conventional manufacturing techniques. The 'third generation' geotextiles were actually designed and developed in new specifications to satisfy the purpose of geotechnical application in particular directionally structured fibres (DSF), directionally oriented structures (DOS) and composite products.[9] The current major use of geotextiles is within the foundation components of load supporting part of a civil engineering structure. Such structure (e.g. building, embankment, dam, canal, road, railway, etc.) transfers its load via its own foundation to the soil mass within and below it. The properties of soils under load are crucial to the stability of any civil engineering structure and to enhance soil stability use of geotextiles gained initial recognition as novel geotechnical materials. More recently, geotextiles have been used to enhance tensile and mechanical properties of civil engineering materials themselves, such as road surfaces and sub-surfaces. However, their behaviour in soils and the characteristics of soils enable geotextile functions and applications to be more readily appreciated.

1.3   Man-made geotextiles

The commercial development of synthetic geotextiles in civil engineering dates back to 1937 although in a meagre way. Large-scale application of such materials started around early 1950s.[10] It all began in the Netherlands in 1953 when the Dutch interest in geotextiles stemmed largely from the need to find innovative construction solutions for use on their massive Delta Works Scheme. In 1970s, the humble figure of synthetic geotextile consumption was around 10.2 million sq. metre while after twenty years in 1990, it was around 420 million sq. metre and in the year 2000, the consumption raised to 760 million sq. meter. These man-made geotextiles mainly constituted of thermoplastic synthetic fibres like polyethylene, polypropylene, polyester, polyamide, polyvinyl chloride and some other petrochemical derivatives witnessed a soaring consumption to 2475 million sq. metre in 2006–07—the growth rate being 10% to 15% since 2000–01. The market for geotextiles is still confined to the USA, Canada, developed countries in Western Europe, Japan, Hong Kong, Germany and Australia. These countries account for nearly 33% of the global consumption. The rate of growth would have been higher had other countries preferred its use during the said period.

1.4   Natural geotextiles

The striking part of the growth of the global geotextile sector is that geotextiles made of natural ingredients like jute, coir, sisal, kenaf, ramie constitute only 5% to 6% of the present global consumption. Admittedly there was not much R&D exercise with natural fibres initially. Behaviour of natural fibres being markedly different from that of man-made geotextiles rigorous research and studies on each of the potential fibres were called for. Once upon a time, woven cotton fabrics were used as an early form of natural geotextile. South Carolina Highways Department used woven cotton fabric smeared with hot bitumen in a series of road construction works and field tests since 1926. These tests continued for nine years and the results for these applications were found encouraging. But cotton fabrics never became properly established as geotextiles, probably due to its limitations to fulfilling the requisite technical requirements. From the foregoing, it is evident that the concept of geotextile as an agent to improve engineering properties of soil originated from the age-old use of natural fibres in making pathways and in controlling erosion in coasts and river banks. While man-made geotextiles dominate in most spheres of application, natural geotextiles made of jute, coir, hemp, sisal, etc. have gradually emerged as potential candidate to match with the former in areas involving erosion, filtration, drainage and separation and they constitute more than 50% of all geotextile applications. In such areas, the desirable factors such as initial strength, drainage capacity and non-clogging filtration ability count very much. In India and other developing countries, use of natural geotextiles is preferred as they are less expensive, available in abundant quantity and are eco-friendly.

1.5 Methods of manufacturing geotextiles

Geotextiles are classified according to their method of manufacturing such as, woven, non-woven, knitted and braided. Woven geotextiles also include fabrics with comparatively large openings (open weave construction), and are made by interlacing of two or more sets of yarns/filaments/tapes or other basic weavable ingredients. The woven fabric is manufactured with two sets of yarns—one is running in the machine direction (called warp yarns) and the other is running across the machine direction (called weft yarns). Non-woven types of geotextiles are manufactured by bonding or interlocking of staple fibres, monofilaments or multi-filaments that are either randomly or specifically oriented. Mechanical, thermal or chemical means and suitable combinations of these bonding methods may be used to achieve the desired bonding or interlocking of fibres. Geogrids, i.e. polymer lattices, are made by perforating extruded polypropylene or HDPE (High Density Polyethylene) sheets. Natural geogrids are uncommon. It is however felt that natural composites may be tried for manufacture of geogrids. Three-dimensional (3D) mats that are manufactured by extrusion of polymers followed by application of hot pressure are brought under the non-woven category (spun bonding). These mats are used mainly for drainage. Application of knitted geotextiles is rare. Knitted fabrics are basically made by inter or intra looping of yarns. Both woven and non-woven geotextiles are permeable. Impermeable geosynthetic composed of one or more synthetic sheets, known as geomembranes, available in the form of planar sheets. Not all impermeable materials or waterproof barriers are however geomembranes. Geomembranes are manufactured from thermoplastics or bituminous products.

1.6 Functions of geotextiles

The major functions of geotextiles imply segregation of two or more layers of materials by preventing their intermixing either dissimilar materials or similar materials with different grading. The placements of a flexible porous textile between dissimilar materials so that integrity and functioning of both materials can remain intact or are improved. The use of textiles in construction and geotechnical engineering can be best explained by their engineering functions which are—confinement, separation, drainage and filtration, reinforcement, protection (for erosion control) and forms.

1.6.1 Confinement

When a load is applied on the aggregate above the sub-soil, there is a tendency on the part of the aggregate to convert static force into dynamic. As a result the configuration of the aggregate is disturbed. To maintain the confinement of the aggregate, there is a need to provide a media between the aggregate and the sub-soil, which should absorb the load in the form of tension and prevent change in the alignment of the aggregate. Such media, apart from other properties should have high friction surface with good transmissivity. Geotextile fabric can meet with the requirement as can be understand from Fig. 1.1. The vertical load induces lateral forces, which spread the aggregate particles and thus lead to local deformation of the fill. Due to frictional interaction and interlocking between the fill material and the geotextiles, the aggregate particles are restrained at the interface between the sub-grade and the fill. The reinforcement can absorb additional shear stresses between sub-grade and fill, which would otherwise apply to the soft sub-grade. This improves the load distribution on the sub-grade and reduces the necessary fill thickness.

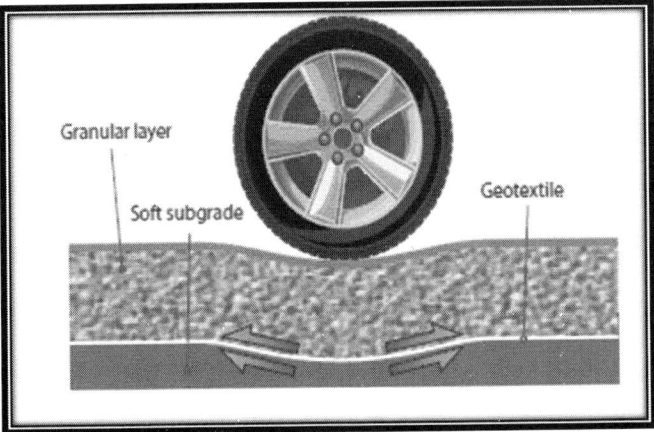

Figure 1.1   Geotextile restraining lateral movement of the material at the bottom of the granular layer.

1.6.2 Separation

Separation is aimed at keeping two dissimilar materials apart and geotextile used for the purpose are required to do the task economically and or better than other available methods. The object of separation by geotextiles is to prevent a well-defined material or rich material from penetrating the sub-grade or the poor soil. The technique is found very useful for separating stone base from soil sub-grade and has been advantageous when applied for designing roads, parking lots, driveways, sidewalks, etc. In such cases, if the separating media of geotextiles is absent, the combined effect of the stone aggregate penetrating the soil sub-grade and the soil sub-grade infiltrating, the stone aggregate decreases the permeability of aggregate to the point where it cannot adequately transport the water that comes to it. The entrapped water obviously leads to the loss of strength, settlement, frost action, potholes and rapid pavement deterioration. Suitable geotextile fabric with good puncture and tear resistance when used as a separator media leads to (a) eliminate the loss of costly aggregate material into sub-soil, (b) prevent upward pumping of sub-soil, (c) eliminate contamination and (d) maintain porosity of different levels. The separation function can be well appreciated by Fig. 1.2 and Fig. 1.3, respectively.

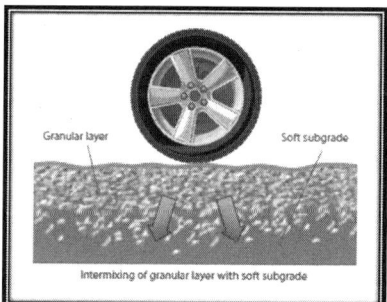 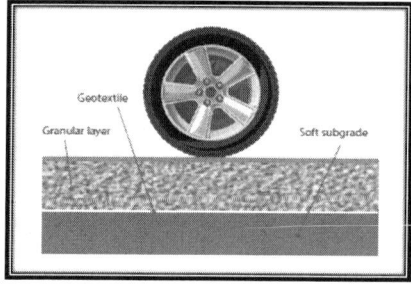

Figure 1.2   (a) Without geotextile—intermixing of granular layer with sub-grade. (b) With geotextile—maintaining boundary between layers.

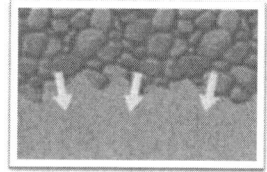 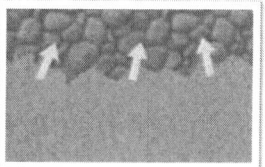 

Figure 1.3   (a) and (b) Without geotextile. (c) With geotextitle.

The strength required for separation in a completed road bed is relatively small compared to that required during construction when normally much higher stresses are induced. There are numerous cases specifically in areas with low bearing capacity (California Bearing Ratio CBR< 3) when the use of the right geotextile, originally intended for separation only, has simplified construction significantly by contributing to the proper stabilization of the soft sub-grade. A prime fabric requirement besides proper drainage properties in these cases is deformation resistance, specifically high tensile modulus. Geotextiles used in road construction are installed in the sub-structure or as a layer separating the sub-structure from the sub-soil. In this respect they are used as separation layer to prevent materials mixing as reinforcement (to give better traffic load distribution), or as a filter. In all three fields (hydraulic structures, road construction and land drainage) geotextiles must satisfy certain requirements for water permeability and soil tightness. Thus, the performance of a geotextile as a separation layer and also its contribution to the road structure as a whole largely depends on the sub-grade type, the size and number of loadings during the service life of the road, the conditions during construction as well as geotextile properties. With regard to the sub-grade the most important factors are the sub-soil grain size and the grain size distribution, the moisture content in relation to plastic limits and the shear strength. The moisture content of the sub-soil or presence of water determines the behaviour of the sub-grade.

1.6.3 Drainage and filtration

The purpose of geotextiles with reference to drainage and filtration is simply to retain soil while allowing the passage of water. When geotextiles are used as drains, the water flow is within the plane of the geotextile itself. At the same time, geotextiles must possess adequate dimensional stability to retain its thickness under pressure. The role of geotextiles in drainage system can be well explained by the fact that the ideal drainage system should remove free water that remains beneath the surface. It has been observed that the life of pavement of Highways or Air Field Pavement, etc. is greatly influenced by the time for which water remains under structural section and its drainage system which is responsible for the removal of the free water which it is fed directly from the stone base course beneath the structure. The use of geotextiles in drainage system can be viewed in Fig. 1.4. In fact, the use of geotextiles in drainage has significant strides into changing the conventional procedure of using graded filters. The outstanding advantages of geotextiles in drainage are—geotextile eliminating the filter sand with the dual media backfill and the need for perforated pipes. In situation where only sand backfill is available it is possible to wrap the drainage pipe with fabric to act as a screening agent. With geotextiles, trench excavation is well reduced and many times the use of geotextiles eliminates need for trench shoring. The geotextile function of filtration involves the movement of liquid through the geotextile itself, i.e. across its manufactured plane. At the same time, the geotextile serves the purpose of retaining the soil on its upstream side. The fabric structure will be such that it will allow the water to pass through and at the same time retain the soil. Another factor is also involved that being a long-term soil-to-geotextile flow compatibility this will not excessively clog during the lifetime of the system. Thus a definition of filtration is the equilibrium soil-to-geotextile system that allows for adequate liquid flow with limited soil loss across the plane of the geotextile over a service lifetime compatible with the application under consideration. Geotextiles are applied as filters in bank protection works, downstream culverts and in bed protection works along canals. In these situations the filter is laid directly on the sub soil and must guarantee the soil tightness of the structure. In order to avoid the buildup of excess pressures under the protection, the geotextile must also be sufficiently water permeable to allow groundwater to pass through.

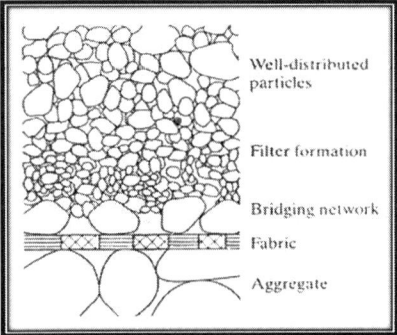 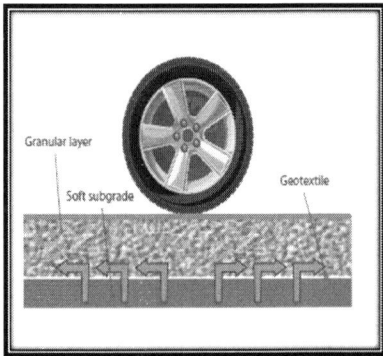

Figure 1.4   Schematic diagram for application of geotextiles as drainage and filtration media where geotextile allows controlled passage of excess pore water from the soft sub-grade.

1.6.4 Reinforcement

The purpose of geotextiles in the reinforcement function is to reinforce the weak sub-grade or sub-soil and this particular application has very wide uses. When the fabric is placed between the sub-grade and the aggregated layer, the level of stresses in the sub-soil decreases which is due to distribution of the vertical load in the form of horizontal shear stress over the fabric. This in turn places the fabric in tension which spreads the loads over a large area and thereby decreases its intensity. As a result, the unit vertical stress on sub-soil is decreased. The decrease in stress means less likelihood of a failure and/or less settlement. However, it is necessary to note that in roadway construction, account must be made for radial stress distribution and the width of the fabric should be wider than the width of the road for allowing anchorage. The functions of geotextiles as a reinforcement material can be better understand from Figs. 1.5, 1.6, 1.7, respectively. Reinforced geotextiles can be used for— (1) Road, (2) Temporary Roads, (3) Pavements, (4) Air Strips, (5) Stabilized Road Slopes, (6) Retaining Walls, (7) Containment Systems, (8) Controlling Reflective Cracking, (9) Fibre or Fabric Reinforced Concrete, etc.

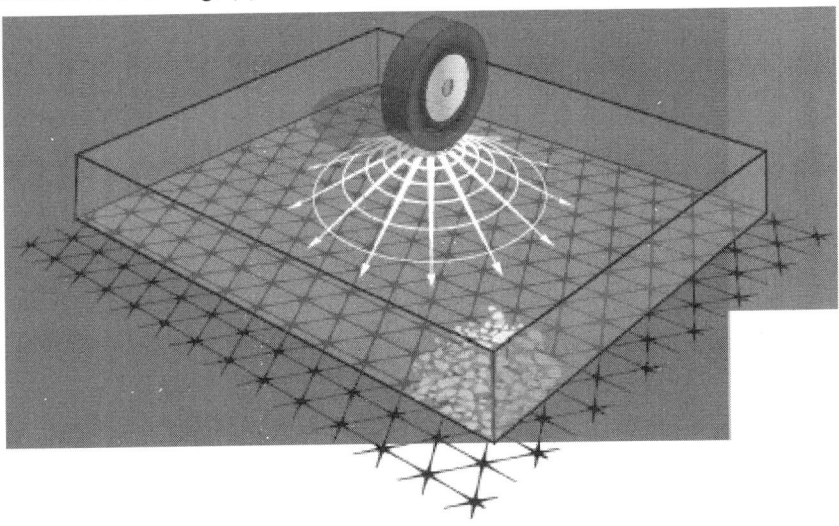

Figure 1.5   Application of geotextile in distributing radial stresses imparted through conventional wheel trafficking.

Source: Tensar International Corporation, 2500 Northwinds Parkway, Alpharetta, Georgia 3009.

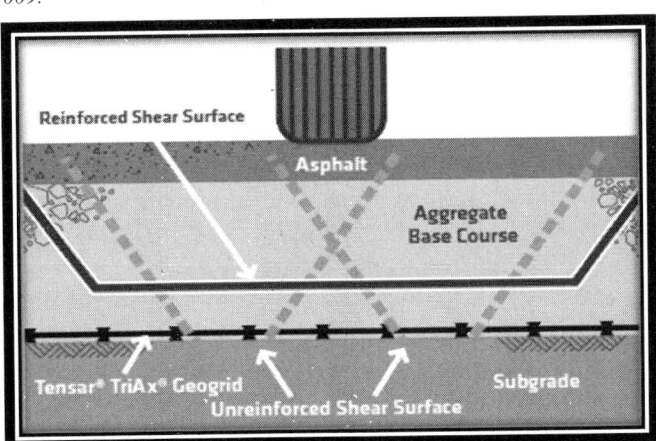

Figure 1.6   Bearing capacity improvement mechanism shown through the inclusion of geotextile at the aggregate base course sub-grade interface.

Source: Tensar International Corporation, 2500 Northwinds Parkway, Alpharetta, Georgia 3009.

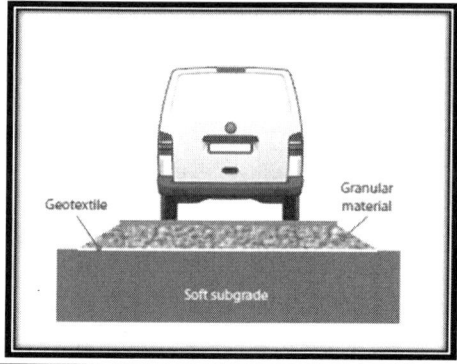

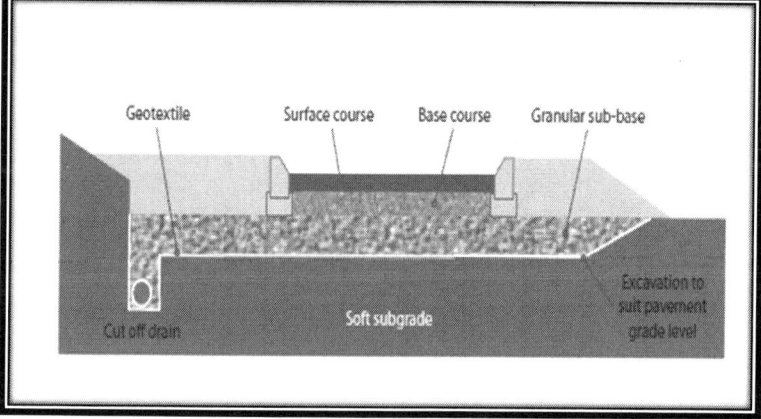

Figure 1.7 Geotextile stabilizing temporary and permanent flexible
pavements.

Source: Terram Limited, Mamhilad, Pontypool, United Kingdom

1.6.5 Protection

Structural erosion can be mainly attributed to forces of wind and water. The
damage caused by a violent storm over a lake or ocean to natural and man-made
objects in its path is a matter known to all. Rainfall too is responsible for soil
erosion. It can erode billions of tons of topsoil. The force of rain hitting the
ground at 20 miles/h can loosen any bare soil. Apart from this, rainfall compacts
the upper layer of the soil thereby reducing its absorptive capacity. This in turn
is responsible for more run-offs resulting in sheet or gully type erosion. Erosion
control is therefore the function of geotextiles, which basically includes the
role as separator and as a drainage layer. There are many applications where
geotextiles can be used for erosion control like it can be used as shore and
coastal beach protection where the fabric acts as a mechanism to hold the soil
in place while allowing for germination of vegetation and weed growth and
this function can be well understood from Fig. 1.8. The fabric can be used as a

boundary material beneath a stone layer or riprap in protecting slopes adjacent to flowing water or in tidal area and artificial seaweed to allow for buildup of natural sediments as well as a slit fencing to block migration of soil from fines being carried by water or by wind.

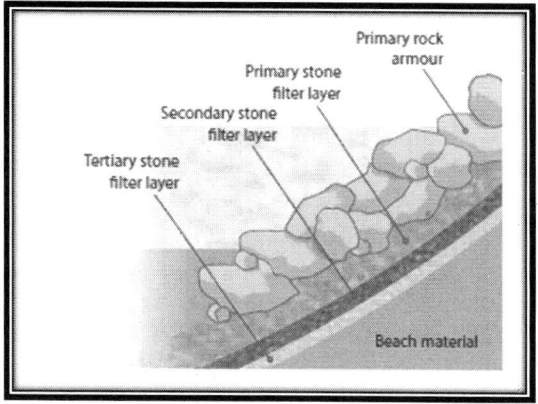

Shoreline with graduated stone filter layer

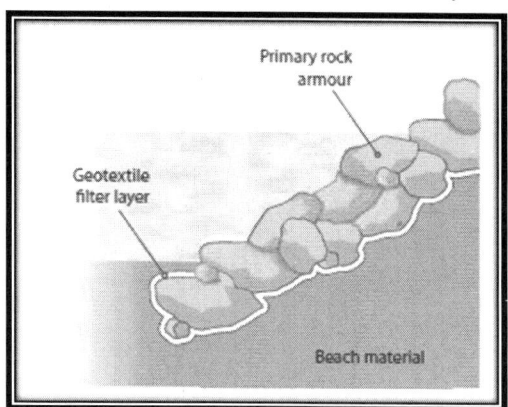

Shoreline with geotextile filter layer

Figure 1.8 Geotextile in coastal beach protection.

1.6.6 Forms

Geotextiles can be used as forms, which are filled with other materials and thereby, conform to the shape and topography of the surface on which they are constructed. Traditional construction methods involve stiff, rigid inflexible framework of concrete, grout and soil fill as forms. It is very easy to visualize that a high permeable fabric can be converted into an enclosure in the desired shape. These enclosures can act as a form of concrete or grout placement. Such systems can work both above the ground and in the water.

1.7   Raw materials for geotextiles

For the purpose of manufacturing geotextiles, the raw material (fibres) should have good resistant to acids, alkalis, oxidizing agent, microorganism, temperature, cold and hot and ultraviolet radiation. To fulfil above requirement, man-made fibres are most suitable raw material for geotextiles. High strength, high modulus, extra creep resistance and chemical inertness are the essential properties of geotextile fibres. Mechanical and dimensional stability and biodegradability is the next greatest importance. Depending on the uses, filament yarns, monofils and tapes of polyamide, polyester, polypropylene, polyethylene and polyacrylic are used in the making of geotextile. Sometimes, spun fibre yarns made from these raw materials are also used. The inertness towards chemicals, low specific gravity, lower cost of volume ratio and easy processability of the polypropylene fibres are probably the main reason behind its popularity. Where the decomposition of the textile is desired in the longer term (e.g. plantation, etc.); natural fibres such as jute, coir, ramie and sisal fibres can be used. Jute and kenaf fibres have much higher relative strength as well as cost competitive as compared to other bast and leaf fibres for producing cost effective technical textiles, which is very much suitable for different geotechnical applications. The other vegetable fibres which are used for manufacturing geotextiles are flax, hemp, abaca, sisal and coir. Table 1.1 shows the different property parameters of some man-made fibres in comparison to the properties of a natural fibre like jute, to understand the suitability of those fibres as geotextile materials with respect to different end uses.

Table 1.1   Comparison of some basic properties of different synthetic fibres and jute fibre, for understanding the suitability of their use as geotextiles

Property Parameters	PET*	PP*	LDPE*	HDPE*	PAr*	Jute
Density (kg m^{-3})	1380	900–910	920–930	940–960	1450	1450
Crystallinity (%)	30–40	60–70	40–55	60–80	90	50–55
Glass transition temperature (T_g) °C	75	−15 to −20	−100	−100	340	–
Melting point (T_m) °C	250–260	160–167	110–120	125–135	550	–

Tensile strength (GN m^{-2})	0.8–1.2	0.6–0.8	0.08–0.25	0.35–0.60	2–2.8	0.43–0.71
Breaking strain (%)	8–15	10–40	20–80	10–45	2.6	1–2
Initial modulus (GN m^{-2})	8–11	5–8	2–4	3–6	50–100	17–28
Moisture regain (%) at 20°C at 65% Rh.	0.4	0	0	0	3	12

PET*: Polyethylene terephthalate; PP*: Polypropylene; LDPE*: Low density polyethylene; HDPE*: High density polyethylene; PAr*: Polyaramid.

1.8 Types of geotextiles

Geotextiles are essentially fabrics having to do with containing, protecting, enhancing, filtering or collecting natural resources both in solid and liquid forms. It is a science of designing certain special fabrics or systems for a wide range of civil engineering application, for the purpose of retaining desirable natural resources or protecting against undesirable elements. Traditionally, textile fabrics are produced by weaving, knitting, non-woven manufacturing or braiding techniques as shown in Fig. 1.9. Geotextiles can be subdivided into several different categories based upon their method of manufacture, i.e. woven, non-woven, knitted, braided or knotted, etc. However, most of the geotextiles used are either woven or non-woven and a few are knitted or knotted types making only a very small contribution to the geotextile market. Woven fabrics are produced by interlacing two sets of spun yarns, tapes, monofilaments (slit-film) or multifilament (slit-film) yarns at right angles to each other. Knitted fabrics are produced by interlooping the same type of yarns, instead of interlacing them. Production of yarns is therefore essential criteria for the manufacture of both woven and knitted fabrics. Non-woven fabrics, which were introduced commercially in late 1920s and early 1930s, are on the other hand made mainly from fibre webs bonded either by mechanical, chemical, thermal or solvent means or a combination thereof. The choice of the fabric for any geotechnical application depends upon the mechanical, chemical and hydraulic properties required for that particular application. Rankilor (1981) and Van Zanten (1986) have discussed in general, the fabric structures currently used for geotextiles and also seaming techniques, which enable their more effective use in the field of civil engineering construction is shown in Fig. 1.10. They have also reported that such structures enable geotextiles to be more effectively engineered and the advantageous properties of anisotropic (bi and multi-axial) fabrics combines with those of more isotropic, high specific volume structures.

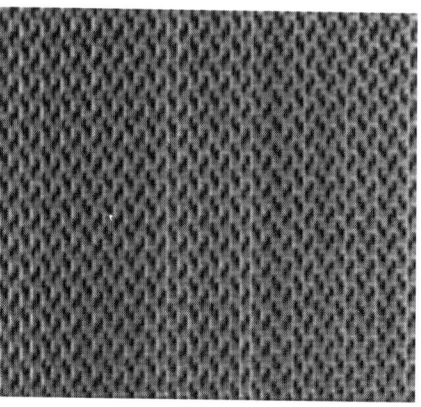

Woven monofilament geotextile

Woven multifilament geotextile

Woven slit film monofilament geotextile

Woven slit film

Nonwoven geotextile

Knitted geotextile

Figure 1.9 Different types of geotextiles.

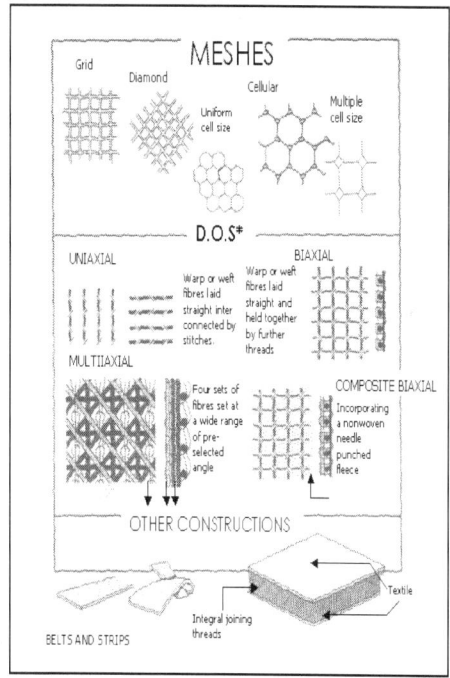

Figure 1.10 Directionally structured fibres (DSF) geotextiles (Rankilor, 1988).

1.8.1 Woven geotextiles

Woven geotextiles were the first to be developed from synthetic fibres and account for approximately a quarter of the geotextiles market in terms of volume. As their name implies, these geotextiles are manufactured using weaving techniques adapted from those used to weave clothing textiles. The weaving process gives these geotextiles their characteristic appearance of two sets of parallel threads, more correctly known as yarns, interlaced at right angles to each other which are shown in Fig. 1.11. The terms 'warp' and 'weft' are used to distinguish between the two different directions of yarn. The yarn running along the length of the loom and hence along the length of the geotextile roll is known as the warp. The yarn running in the transverse direction, across the width of both the loom and the geotextile roll, is known as the weft. Although this form of construction may appear quite simple there are many different variations on the basic concept, giving a fairly wide range of different forms of woven geotextile. Most commonly used weaves in the formation of fabrics for use as geotextiles are—plain weave, basket weave, twill weave, etc. In the plain weave, which is the simplest weave pattern, both the warp and the weft yarns follow an over one, under one sequence as illustrated in Fig. 1.11(a). Twill weaves have a characteristics diagonal pattern to the exposed lengths of warp and weft as shown in Fig. 1.11(b). The warp yarns jumping a greater number of weft yarns and vice-versa produce this. A satin weave is achieved by extending the weft and warp jump distance even further and this produces a fabric where the exposed surface is almost entirely warp yarns on one side of the fabric and almost entirely weft yarns on the other. The simplest satin weave is an over five, under one, over five, under one sequence as shown in Fig. 1.11(c). To date, this weave has been limited to the clothing industry and has not been applied to geotextiles. In contrast, the other two basic weave patterns, plain and twill, are widely used in woven geotextiles, together with many different derivatives of these patterns. Similarly, different examples of more complex weave pattern such as compound weave, plain leno weave and triaxial weave are shown in Figs. 11(d), (e) and (f), respectively. Leno weave has a particularly high open area and is formed by twisting pairs of warp yarns between each weft yarn. Its main geotextile applications are in the control of rainfall erosion and within composite geotextiles where it may be used as a scrim fabric. The normal arrangement of two sets of threads in only two directions gives woven geotextiles a distinctly anisotropic strength and stiffness. It is technically possible to weave the weft at a skew angle, sometimes incorporating more than one weft yarn direction on special looms. An example of such a weave is the triaxial weave pattern shown in Fig. 1.11(f). Its main advantage is that the three-way pattern of warp and weft yarns gives it less anisotropic strength properties than conventional biaxial weave patterns. The triaxial weave pattern also achieves a very large open area without making the weave slack at the warp/weft crossover points. This geotextile property could, for instance, be usefully applied in the control of rainfall erosion.

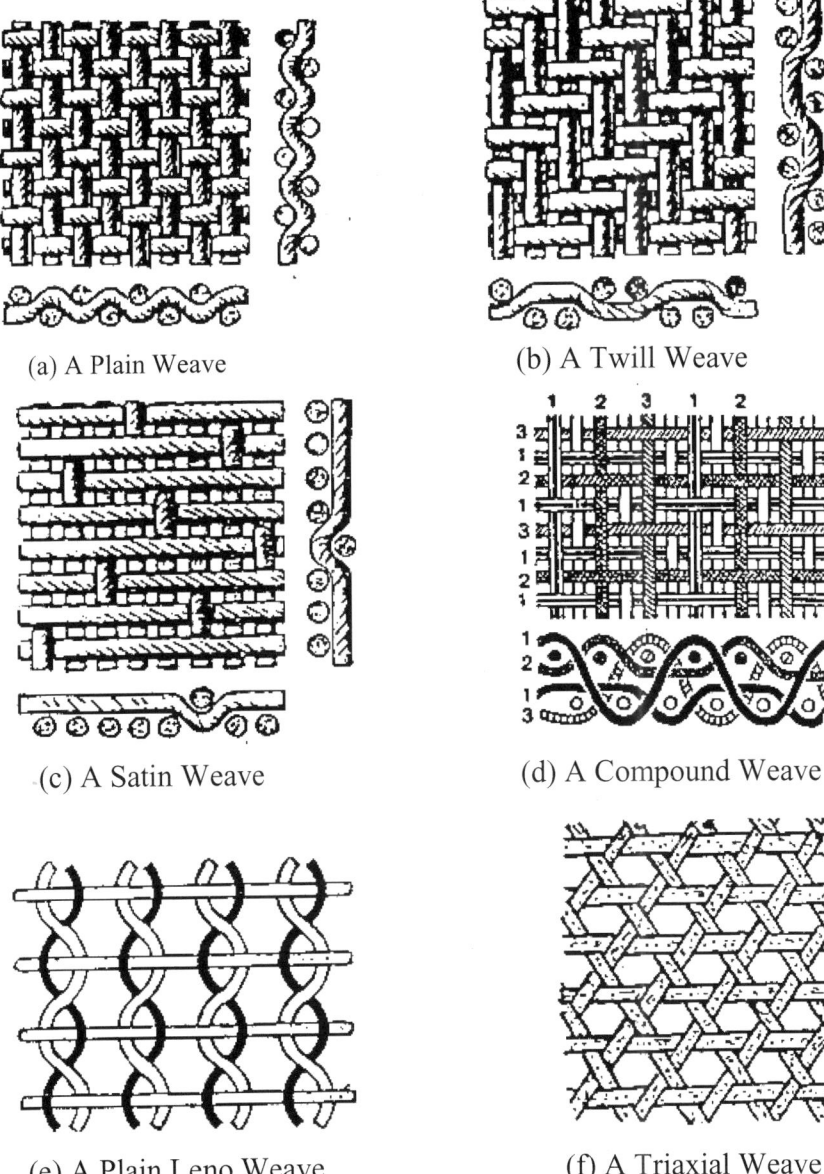

(a) A Plain Weave   (b) A Twill Weave

(c) A Satin Weave   (d) A Compound Weave

(e) A Plain Leno Weave   (f) A Triaxial Weave

Figure 1.11   Schematic view of some woven fabric construction.

1.8.2 Non-woven geotextiles

Non-woven technical textiles are widely regarded as the most thriving and fast changing sector of the global textile industry. The field of non-woven is very versatile, spanning applications from geotextiles to health care and scales of size from structural components for roads and buildings to nanofibres used in filtration that are seventy five times smaller in diameter than a human hair. Innovation in new processes, products and their different applications are expanding non-traditional end uses for both existing and new non-woven textile products. In short, non-woven technical textiles are about function rather than fashion. Initial demand of geotextiles was met by woven fabrics, the spurt in demand of geotextiles was observed only on introduction of non-woven into geotextiles market. The earliest non-woven geotextiles were made of continuous filament needled polyester and spun bonded polypropylene. According to ASTM, definition of non-woven is 'a textile structure consisting of a web of matt of fibres held together by bonding material'. However, in the light of modern non-woven technology the definition of ASTM does not hold good too. Because, nowadays, needle punched non-woven fabrics are quite promising in different civil engineering construction. To accommodate these fabrics into the fold of non-wovens, the definition given by Patterson is rather general and more acceptable. According to him, 'Nonwoven fabrics are essentially two-dimensional assemblage of textile type fibres, held together by means of an additive of self bonding material, resulting in a mechanically stable self-supporting, flexible web-like structure'. The recent definition of non-woven fabric given by INDA is as follows, 'Structures made by bonding fibres, yarns or filaments by mechanical, thermal, chemical and/or solvent means. Structures include but are not limited to air laid structures, bonded continuous filament structures (spun bonded), carded structures, fibre reinforced structures, fibrillated film structures, laminated and other composite structures, needle-punched structures, point bonded structures, scrim reinforced structures and water formed long fibre structures'. The ideal non-woven fabric structures as assumed by Hearle is fibres with high crimp through unbalanced forces, web with the right long range fibre paths and three-dimensional structure, a combination of interlacing and point bonding and right overall fabric mechanics. Non-woven fabrics account for 75% to 80% of the fabrics used as geotextiles and other technical applications and most of these fabric types are needle punched and spunlaid continuous filament fabrics in nature.

1.8.2.1 Types of non-woven fabrics

The basic idea of the non-woven technology is to produce fabrics directly from fibres without intermediate stage of fibre conversion in form of yarns. However, in some non-woven structures, yarns or other materials may be used to either bond or reinforce the structure. Therefore, the ranges of possible combinations to form a non-woven are many and varied with continuous additions of new technologies. Textile structures classified as non-woven fabrics are summarized in Fig. 1.12. Non-woven geotextiles are normally made from extruded continuous filaments or staple fibre webs, which are bonded using mechanical or thermal means. Needle punched and thermally bonded non-wovens are now being extensively used as constructional material in conjunction with other geotechnical materials such as soil and rock in applications of civil engineering nature. Fabrics produced by either needle punching or thermal bonding of continuous filament webs are commonly known as spun bonded or spunlaid fabrics. A diagrammatic view of different types of non-woven fabrics, which are mostly used in the field of civil engineering construction is shown in Figs. 1.13(a), (b) and (c), respectively.

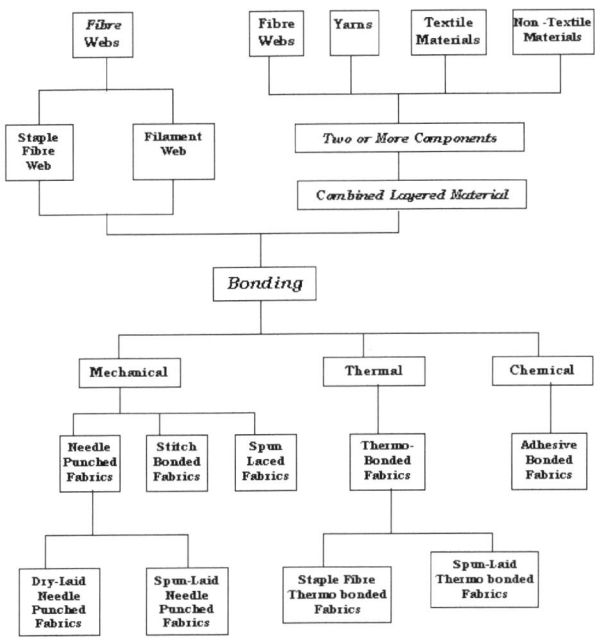

Figure 1.12 Classification of non-woven fabrics highlighting the methods and fabrics of importance as geotextiles.

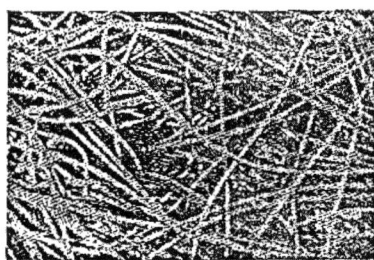 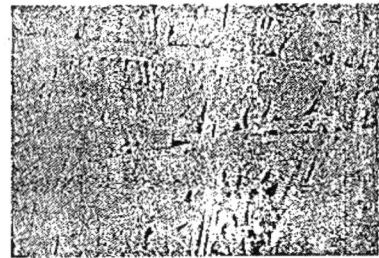

(a) Photomicrograph of a continuous filament needled nonwoven geotextile-surface view

(b) Photomicrograph of a spun bonded nonwoven geotextile-surface view

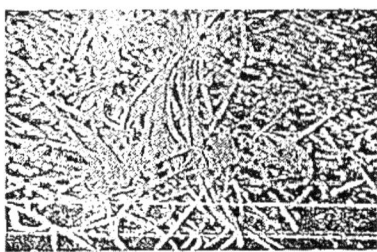

(c) Section of a needle bonded geotextile with porosity of 93%.

Figure 1.13   Photographic view of some non-woven fabric construction.

1.9   Market analysis of geotextiles

Global geotextile market size was estimated to be USD 4.11 billion in 2013. These fabrics are manufactured using synthetic or natural fibers and classified as knitted, woven or nonwoven based on specific applications. Longer life span and cost effectiveness in comparison with other materials and growing environment concern is expected to increase demand.

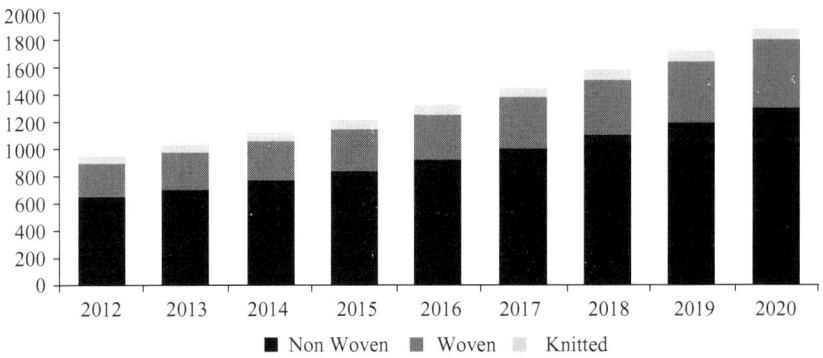

Figure 1.14   Asia Pacific geotextile market volume by product, 2012–2020 (million square metres).

Source: http://www.grandviewresearch.com/industry-analysis/geotextiles-industry, Published: April 2014/ISBN: 978-1-68038-023-1.

Non-woven fabrics dominated the market with 65.3% of global volume in 2013, owing to unique properties such as absorbency, liquid repellency, and mechanical strength. Non-woven textile manufacturing does not include conversion of fibre into yarns, thus reducing overall costs. Key industry players including Royal TenCate, GSE Holdings and Agru America have non-woven fabrics in their product portfolio. Infrastructural developments in Asian countries such as China and India are expected to improve demand for non-woven fabrics over the forecast period.

Woven geotextile accounted for 26.5% of total revenue in 2013. High prices and limited availability as compared to non-woven textiles are factors which will restrain woven growth in future. In addition, heavier weight of woven coupled with limited ranges of fabric availability for woven textiles act as factor inhibiting growth of woven products

References

1. Giroud, J.P. and Perfetti, J., Use of fabrics in geotechnics. In: Proc. Int. Conf., Paris, France, Vol. 2, pp. 345–352, 1977.

2. ASTM-D-4439-02, Geosynthetics, Annual Book of ASTM Standards, D-35 on Geosynthetics, Section 4, Vol. 04.13, pp. 1–18, 2003.

3. Ingold, T.S., The Geotextiles and Geomembranes Manual, first edition, Elsevier Science Publishers Ltd., Oxford, pp. 1–3, 121–127, 1994.

4. Venkatappa Rao, G., Banerjee, P.K., Shahu, J.T., Ramana, G.V., Geosynthetics–New Horizons, Asian Books Pvt. Ltd., New Delhi, pp. 1–8, 1994.

5. Sanyal. T., Proc. Int. Jute Symposium on Environmental Applications of Jute Geotextiles, Kolkata, West Bengal, India, pp. 143–160, 2003.

6. Giroud, J. P. and Han, J., Design method for geogrid-reinforced unpaved roads. I: Development of design method. Journal of Geotechnical and Geoenvironmental Engineering, 130(8), 775–786, 2004.

7. Giroud, J. P., and Noiray, L., Geotextile-reinforced unpaved road design. Journal of Geotechnical Engineering, 107(9), 1233–1254, 1981.

8. Horrocks, A. R., The Durability of Geotextiles, Eurotex, Bolton, UK, pp. 1–3, 1992.

9. Rankilor, P.R. and Raz, S., The fundamental definition and classification of warp knitted D.S.F. geotextiles for civil engineering end uses, 1988.

10. Ramaswamy, S.D. and Aziz, M.A., Jute geotextiles for roads. In: Proc. Int. Workshop on Geotextiles, Bangalore, India, 22–29th November, pp. 259–266, 1989.

Jute and jute geotextile (JGT)

2.1 Jute fibre

Before we discuss on Jute Geotextiles (JGT), it is relevant to know the properties, features and characteristics about jute, jute fibre and jute yarn. Jute is a very old agricultural product cultivated mostly in the Gangetic Delta. Its leaves were consumed as vegetable and used as a household herbal remedy.[1] It was however after setting up of jute mills in the vicinity of Kolkata in the mid-nineteenth century that its cultivation gained importance and systematized. The spurt in use of jute fibre for making of sacks led to improvement of its method of cultivation and special extraction of fibre (retting) followed by manufacture of fabrics with its yarns. Jute is the common name given to the fibre extracted from the stems of plants belonging to the botanical genus *Corchorus*.[2] The genus *Corchorus* (family Tiliaceae) includes about 40 species distributed throughout the tropics. Of all the species of *Corchorus*, *C. capsularis* Linn. and *C. olitorius* Linn. were selected by man in the wake of civilization as economic plants for extraction of coarser variety of bast fibre that is commercially important; while the other species are found wild in nature. *C. capsularis* is known as 'White' jute and *C. olitorius* as 'Tossa' jute.[3] Jute plant is known to be annually agro-renewable crops (Indian Grass) mostly cultivated in the Bengal Delta since 800 BC. In India, it is the most important commercial agro-renewable fibre crop of nature next to cotton.[4] India, alone, accounts for about 65% of world production of jute and allied fibres.[5, 6] It thrives in hot and humid climate, especially in areas where rainfall is abundant. It grows up to about 3 metres in height usually and matures within four to six months. Jute fibres are extracted from the thin bark and woody core of the plant by a special manual process known as '*retting*'.[1]

Organized cultivation of jute took off probably in the beginning of nineteenth century, after samples of coarser jute fibres extracted from its plants (after water retting, Figs. 2.2 and 2.3) were sent to England for tests of commercial spinnability for yarn making by the East India Company in 1791.[7] After satisfactory feedback of the said tests, jute fibre was accepted as a coarser variety of commercial fibre for making coarser yarns which in turn can be woven to sacking, wrapping and backing fabrics of jute, following the process recommended by the experts of Dundee in Scotland, which was the major textile processing, research and training centre in that region. Suitability of jute fibres in making of sacks was appreciated and its demand soared particularly in the period between two major wars, Crimean War (1854–1856) and American Civil War (1860–1865).[8, 9]

Figure 2.1   Jute plant.

Figure 2.2   Submersion of jute plants for retting.

Figure 2.3   Drying of jute fibre.

Uses of jute as wrapping hessian fabric, carpet-backing fabric had been popularized later. However, jute hessian was first tried in a road as a reinforcing material at Dundee, Scotland in 1920 and later on in Strand Road, Kolkata, India in 1934, both appreciated more or less satisfactory results. However, for unknown reasons, these trials unfortunately were not monitored, followed up and continued in right perspective of the potential of jute in road construction and this potential remained unexplored. The USA started using open weave jute geotextiles (under brand names of 'Soil Saver', 'Antiwash', etc.) principally for the control of slope erosion, which, till date, remains as one of the major exportable jute product of India.

2.2 Properties of jute fibre

Jute is a lignocellulosic and multicellular natural bast fibre produced mainly in India, Bangladesh, China and Thailand, which possess conducive agro-climatic and socio-economic conditions suitable for production of jute.[10] The average linear density of single jute filament lies between 1.3 and 2.6 tex for white jute and 1.8 and 4.0 tex for tossa jute with normal distribution.[11] Coarseness of jute has some role in determining the strength of jute fibre. Coarse fibres are usually stronger. Jute fibre is usually strong with low extensibility.[12] It has a tenacity range of 4.2–6.3 gf/denier.[13] Tenacity varies with the length of the fibre. Elongation-at-break of jute fibre is between 1.0% and 1.8%.[14] Tossa jute is stronger than white jute. Jute fibre breaks within elastic limit and is resilient which is evident from its recovery to the extent of 75% even when strained quite a bit (1.5%).[15] Its flexural rigidity and torsional rigidity are high compared to cotton and wool.[16] Presence of hemicellulose in jute fibres makes it hygroscopic, second only to wool.[17] Tossa jute is slightly more hygroscopic than white jute.[18] Jute fibres swell on absorption of water. Lateral (cross sectional) swelling of jute fibres (about 45%) far exceeds its longitudinal swelling (0.4%).[19] During the process of addition of water, tenacity of jute fibre increases at the initial stages up to the relative humidity of 20% which does not vary for most of the period of water addition thereafter, but exhibits a downward trend when the relative humidity exceeds 80% or so.[20] This phenomenon implies decrease in flexural and torsional rigidity of jute fibre when moisture absorption exceeds a limit.[21] Jute is not thermoplastic like other natural fibres. Charring and burning of jute fibres without melting is a feature which jute fibre possesses. Due to high specific heat, jute has thermal insulation properties. Ignition temperature of jute is of the order of 193°C.[22] Long exposure of jute fibre to hot ambience reduces its strength. Dry jute is a good resistant to electricity, but it loses its property of electrical resistance appreciably when moist. Dielectric constant of jute is 2.8 kHz when dry, 2.4 kHz at 65% RH and 3.6 kHz at 100% RH. Coefficient of friction of jute fibre is usually 0.54 for white jute and 0.45 for tossa variety.[23] Moisture content in jute helps increase its frictional property.[24] Treatment of natural fibres with alkali has been studied by a number of researchers. It has been observed that natural fibre surfaces being rich in hydroxyl groups, provides suitability to chemical modification by way of treatment.[25] Alkali treatment is however found to weaken natural fibre. Mohanty et al. studied the effects of treatment on two varieties of jute fabric—hessian cloth and carpet backing cloth—with alkali and other chemical processes like de-waxing and grafting.[26] The results reveal that alkali-treated jute fabrics possess higher tensile and bending strength than de-waxed jute fabrics probably due to improvement of the adhesive characteristics of the fabric on treatment. Jute fibre develops crimps like wool when treated with strong alkali (18%) due to irregular swelling.[27] The process is known woollenization. Acids affect jute fibre adversely and weaken them.[28] Strong acids may destroy jute fibre. Inorganic acids affect jute fibres worse than organic acids. Bleaching agents also affect jute fibres. Table 2.1 illustrates the ranges and averages of different physical and related properties of jute fibre. Table 2.2 shows a comparative data on important physical properties of jute and some other fibres revealing the relative position of jute fibre with respect to other fibres at a glance.

Table 2.1   Different physical and related properties of jute fibre

	Properties	Range
1.	Length of ultimate cell (mm)	0.75–6.0
2.	Width of ultimate cell ($\times 10^{-3}$) (mm)	5.0–25.0
3. (a)	Fibre fineness (linear density) (tex)	0.90–3.50
3. (b)	Fibre fineness (linear density) (den)	8.00–31.0
4.	Fibre density (g/cm^3)	1.45–1.52
5.	Single fibre tenacity (g/den)	3.0–6.0
6.	Breaking elongation (%)	0.8–2.0
7.	Work of rupture (g-cm/cm-den)	0.30
8.	Modulus of torsional rigidity (dyne/cm^2) $\times 10^{10}$	0.25–1.30
9.	Initial modulus (g/den) (modulus at a 0.1% extension)	130–220
10.	Young's modulus (dyne/cm^2) $\times 10^{11}$	0.86–1.94
11.	Bundle tenacity (g/den)	2.2–4.0
12.	Moisture regain (%) at 65% RH and at 27°C	12.5–13.8
13. (a)	Refractive index (parallel to fibre axis)	1.577

13. (b)	(Perpendicular to fibre axis)	1.536
13. (c)	Bi-refringence (double refraction)	+0.041
14. (a)	Swelling in water – diametrical (%)	20.0–22.0
14. (b)	Areawise (%)	40.0–50.0
15.	Stiffness index (g/den)	300–400
16.	Specific heat (cal/g/°C)	0.324
17.	Dielectric constant (at 50 Hz)	2.2–7.2
18.	Insulation resistance (Ω)	10^{10}–10^{17}
19.	Heat of combustion (J/g)	16.0–17.0
20.	Specific internal surface (m^2/g)	10–200
21.	Crystallinity (%)	52–60

Table 2.2 Physical and related properties of jute and some other textile fibres[29]

Fibres	Fibre density (g/cm³)	Fibre fineness (linear density) (den**)	Crys-tal-linity (%)	Tenacity (g/den***)	Break-ing ex-tension (%)	Stiff-ness index (g/den)	Mois-ture regain (%)
Jute	1.48	18	50–55	2.7–5.5 (190)*	0.8–2.0	350	13.75
Flax	1.50	11.7	65–70	5.0–6.5(203)*	1.5–2.5	300	12.00
Ramie	1.55	6.3	60–65	6.0–8.0 (55.5)*	4.0–5.0	159	8.00
Cotton	1.55	1.8	55–60	3.5–3.8 (55.5)*	7.02–8.0	44	8.50
Viscose rayon	1.50	1–4	35–45	2.2–3.0 (45.5)*	15–20	20	12.00
Polyes-ter	1.34	1–4	45–50	3.8–6.0 (111)*	20–55	14	00.40

*Data in the parentheses indicate the initial modulus values for the corresponding fibres in g/den.

**denier (den) = 9 × tex; (g/tex) = 9 × (g/den).

***cN/tex = 8.83 × (g/den).

2.3 Chemical composition and structure of jute fibre

Jute, chemically being lignocellulosic in nature, comprises mainly polysaccharides and lignin.[30] Although a number of minor components, such as pectin inorganic salts, nitrogenous substance, colouring mater, wax, etc. are found in it. The detail of chemical composition of the jute fibre is given in Table 2.3.

Table 2.3 Average chemical composition (in percent of bone dry weight of the fibre) of jute *C. capsularis* (white), *C. olitorius* (tossa)[60]

Constituent	*Capsularis* (white) jute	*Olitorius* (tossa) jute
Cellulose*	60.0–63.0	58.0–59.0
Lignin	12.0–13.0	13.0–14.0
Hemicellulose**	21.0–24.0	22.0–25.0
Fats and waxes	0.4–1.0	0.4–0.9
Proteins or nitrogenous matter, etc. (% nitrogen × 6.25)	0.8–1.87	0.8–1.56
Pectins	0.2–0.5	0.2–0.5
Mineral matter (ash)	0.7–1.2	0.5–1.2

*Major constituents of jute-cellulose include glucosan (55.0–59.0%), xylan (1.8–3.0%) and polyuronide (0.8–1.4%).

**Major constituents of jute-hemicellulose include xylan or pentosan (15.5–16.5%), hexosan (2.0–4.0%), polyuronide (3.0–5.0%) and acetyl content (3.0–3.8%).

Distribution of lignin is not uniform in jute fibre. There is concentration of cellulose and hemicellulose in the primary and secondary walls. Each fibre element of a raw jute stem incidentally comprises 5 to 30 polygonal unit cells (ultimate cells) bonded together, each having a central lumen, primary/secondary cell walls and middle lamella. These ultimate cells are on average 2.5 mm long with tapered ends and a wider middle (18 μm approx.).[31] The diameter of the ultimate cell ranges between 0.0051 and 0.0254 mm. The layer of the bonding substance between the ultimate cells is called middle lamella. Figure 2.4 shows microstructure of jute fibre. The cell wall resembles a hollow tube having two different walls—one primary or elementary layer and a thicker secondary wall composed of microfibrils besides a lumen. Lumen is like an open conduit running through the centre of the microfibril. Each layer contains cellulose embedded in a matrix of hemicellulose and lumen. Hemicellulose is composed of highly branched polysaccharides including glucose, mannose, galactose, xylose, etc. Lignin contains hydrocarbon polymers found around fibres. The layer with microfibril is the thickest and determines the overall properties of the fibre. The microfibril layer is rich in cellulose molecules.

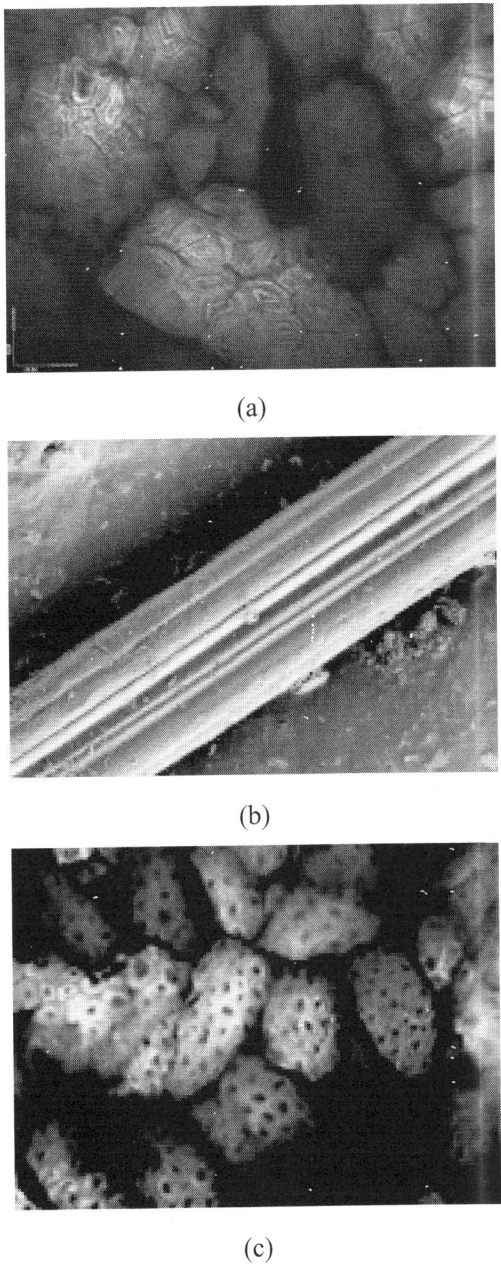

(a)

(b)

(c)

Figure 2.4 (a) Multicellular structure of jute fibre, (b) longitudinal view (5000× magnification) and (c) cross-section (180× magnification) of jute fibre

Source: M. Sfiligoj Smole, S. Hribernik, K. Stana Kleinschek and T. Kreže, Advances in Agrophysical Research, Chapter 15: Plant Fibres for Textile and Technical Applications, http://dx.doi.org/10.5772/52372, pp. 377.

2.4   Grading of jute fibre

Jute fibre extracted by the retting process from the bast of the parent plant comes in the form of long mesh of interconnecting fibres commonly known as jute reed. The jute reed is usually 6–15 feet long. Typical yield of jute fibre based on weight of stem from which it is derived is about 6%. The top of jute reed is thinner than the root. The reeds are then split-opened in carding machine into the component fibres called the spinner's fibre. The Bureau of Indian Standards (BIS) in its publication IS:271-2003 has recommended grading of raw jute based on the fibre characteristics. The characteristics are—bundle strength, fibre fineness, reed length and root content, freedom from defects, bulk density, colour and lustre. There are sub-features to these characteristics. Depending upon these six fibre quality attributes white (W) and tossa (TD) jute are graded into eight varieties, viz. W1 or TD1, W2 or TD2, W3 or TD3 in this way it continues till W8 or TD8, in descending order of quality.[32] Grading is done giving due weights to physical attributes of jute fibre. Maximum stress is given on fibre strength and root content at the time of evaluation of the grading.

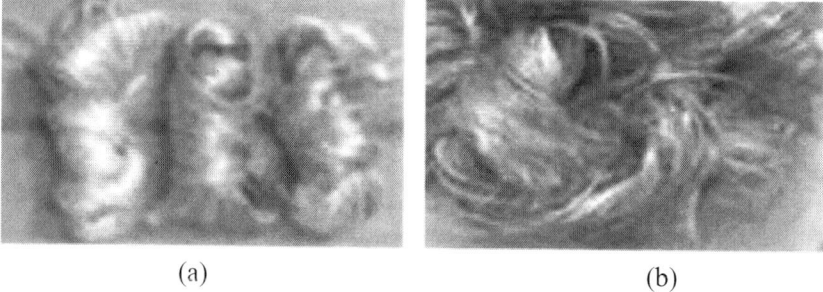

(a)   (b)

Figure 2.5   (a) Tossa jute (golden, red, grey, mixed in colour), (b) White jute (white, creamy, straw in colour).

2.5   Jute yarns

Jute fibres are made into jute yarns[1] which constitute the basic ingredient of any jute textile including jute geotextile. Understandably, good fibre quality ensures good yarns and consequently good textiles. Technically, any yarn is an assembly of fibres and/or filaments either in twisted or untwisted form having its length substantially higher than its diameter or width.[33] Yarns may be spun from staple fibres or may be made directly from continuous filaments.[18] Spun yarns may be made out of more than one type of fibres. The spinning system plays an important role in determining the yarn quality.[34] Structurally, yarns made of continuous filaments are simpler.[35] The difference in yarn type is based on the number of fibres/filaments, irregular features, diameter, hairiness, packing density and the amount of twist exerted in their making. The basic operations for conversion of jute fibre into yarns are opening, cleaning and mixing, formation of slivers, thinning of slivers, parallelization of fibres and packaging.[36]

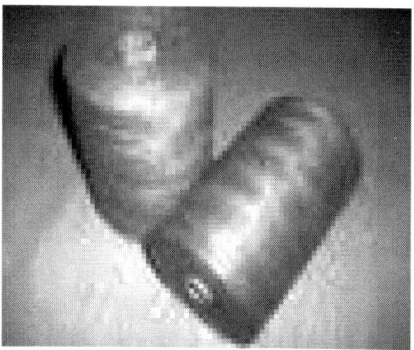

Figure 2.6   Jute yarns.

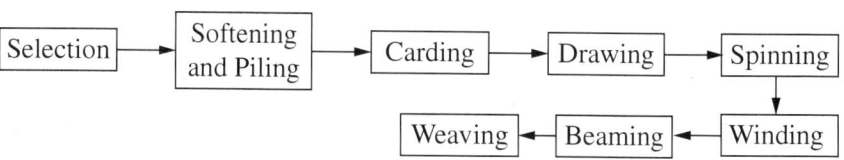

Figure 2.7   Flowchart of processing of jute yarns.

2.5.1 Processing of jute yarns

The processing of jute yarns comprise selection of the appropriate fibre, softening and piling, de-knotting and cleaning, formation of sliver (continuous strands), reduction of sliver linear density, parallelization of fibres, packaging.[37] After formation of sliver, sometimes fibres of different quality are blended. Generally, jute mills follow two spinning systems. In Long Staple System, slivers produced from long jute reeds are processed in the spinning machine. In the second system, known as Short Staple System, slivers of short jute yarns are fed to the spinning machine. Spinning is usually done following Long Staple System. The essential features of spinning process are drafting and winding.[1] Drafting is the process to reduce the bulk and weight of sliver and parallelizing the fibrous components of yarns. Winding is the operation to transfer yarn from one form of packaging to another. The processing of jute yarns can be shown at a glance by the flowchart shown in Fig. 2.7.

2.5.2 Jute yarn quality

The process of weaving of jute fabrics is directly linked with average strength of yarn. Unevenness, hairiness and other imperfections in the jute yarns need to be effectively controlled. Unevenness and hairiness of yarns may affect the fabric porometry, i.e. its apparent opening size. It is also difficult to adhere to the specified weight of jute woven fabric if imperfection in yarn quality creeps in. Irregularity in jute yarns cannot be totally avoided due to random disposition of and defects in fibre.[17] There are methods of measuring such irregularity. Drafting wave (wave-like variation in yarn thickness when subjected to attenuation by rollers) also causes yarn irregularity.[38] Hairiness in yarns has direct relation with the number of fibres in a yarn. Irregularities may be short term and long term.[39] Variation of mass or diameter between successive short yarn segments is called short-term yarn irregularity. For long segments (>100 times the average fibre length) the irregularity is long term.[40] Irregularities in yarns are measurable and these adversely affect the yarn strength. Irregularities in yarn also affect flexural and torsional rigidity.[41] In case of fibres, low flexural and torsional rigidity renders flexibility to yarn formation. Higher flexural rigidity of fibres and also yarns has effects on spinnability and fabric structure. In case of jute, moisture in the right proportion can help reduce both flexural rigidity and torsional rigidity. Yarn twist, if not imparted properly, may adversely affect some of the fabric characteristics especially drapability, abrasion resistance and tensile strength.[42] It requires specific attention to predetermine the optimum twist that yields the highest strength to a yarn. Longer or finer fibres require lower optimum twist while finer yarns need higher optimum twist than coarser ones. The lower the optimum twist, the better the fabric qualitatively. In addition to fibre quality, process control is a critical factor in ensuring yarn quality. Irregularity of a yarn depends on the fibre length and its distribution. Finer fibres usually produce yarns with regular features. The fibre length and length distribution are critical in respect of spinning a good quality yarn. Long fibres are needed for yarns to be strong. Processing factors also influence yarn quality. Softening, carding, drawing, sliver coherence, spinning, winding, twisting should be carried out with care and caution. There are specified systems of these operations that ensure retention of the right yarn quality. Higher moisture content reduces the flexural and torsional rigidity of jute fibre but enhances its toughness. There is thus need for optimizing the moisture content.

2.6   Jute geotextile

Traditionally, jute has been known as a material for bulk flexible packaging in the form of sacks. Jute industry, perhaps the oldest surviving agro-industry in the world on which more than four million people depend directly and indirectly in India, has thrived on this particular product. With intrusion of polymeric fibre in this sector, jute industry has been on the lookout for new avenues for survival. JGT is an effort in this direction. Special features of jute fibre have been gainfully utilized in development of jute geotextile matching its synthetic counterpart technically in all its features except of course durability and tensile strength.[43] Synthetic geotextile is a proven engineering material, but its non-biodegradability and long life pose environmental apprehensions as to its eco-compatibility.[44] Growing environmental concerns have prompted scientists/technologists to look for natural, biodegradable alternatives for geotechnical applications. Jute with its unique features fits in with this aim. High tensile strength of geotextiles is an occasional and application-specific requirement in special cases.[45] The application area of geotextiles is increasing continuously with the development of modern scientific and technological innovation and jute geotextiles are emerging technical textiles in geotechnical and bio-engineering fields.[46] These are fabricated by both man-made and natural fibres with different designs, shapes, sizes and compositions according to site specific conditions and functional needs. These are a group of commodities, which are used for solving problems related to geotechnical, bio-engineering, agronomic and horticultural requirements by way of consolidation, filtration, separation and management of soil along with agricultural mulching. In respect to their physical, mechanical, hydrological properties, natural geotextiles particularly jute geotextiles are getting increasing acceptability due to their environmental complementary support.[47] However, non-environment friendly nature for most of synthetic fibres, particularly polyolefines stand against their continued use. Therefore, the growing consciousness regarding environment preservation has changed the situation in the recent years. Some major plus points regarding jute in this context are include agro-origin, annually renewability, soil friendly organic criteria and complete biodegradable nature, eco-compatibility and improvement of soil fertility and texture. Therefore, use of jute geotextile, as a geotextile material can be capable for overall survival of old-age jute industry as well as jute cultivators. A vast range of diversified jute products along with jute geotextile can be manufactured through vertical and horizontal modification of existing technology and machineries.[48] Jute geotextiles can be described as natural fibre materials used for civil engineering purpose to meet technical as well as functional requirements for soil related problems.[49] It is an engineering fabric which, when placed in or on soil, helps to improve its engineering performance against extraneous loads by acting as a change agent or a catalyst. Independent research in the laboratory and field trials has shown jute geotextiles to be technically 'fit for purpose', especially in the fields of soil erosion control and vegetation management. There is also potential use of these products in the stabilization of rural earth roads. Functionally there is no difference between man-made geotextile and jute geotextile, though life of man-made geotextile is much longer. But as geotextile acts as a change agent for a limited initial period, shorter life of jute geotextile is not a technical deterrent. Interestingly, it has also revealed from the laboratory studies that the rate of gain in strength in soil is compensated by the rate of degradation of jute geotextile. Extensive laboratory studies and field trials with jute geotextile have substantiated its efficacy in addressing a number of soil-related problems in the field of civil engineering construction.

The earliest example of jute woven fabric geotextiles for sub-grade support was in the construction of a highway in Aberdeen in the 1930s.[50, 51] Jute mesh was probably first used in erosion control in USA in the early 1930s, where soil conservation was said to have taken a modified form of jute mesh used to wrap bales of cotton and laid on slopes to prevent wash-off from newly seeded grounds. These types of construction however are more comparable to reinforced concrete than today's reinforced earth techniques, because of the rigid way in which stress was transferred to the tensile elements and the 'cemented' nature of the fill. Ramaswamy and Aziz[44] reported on the economical design and construction of haul roads on poor sub-grades with jute fabrics. It is reported that jute geotextile had also been used for mine spoil stabilization, hill slope protection and sand dune stabilization in 1987 and 1988. It has been observed that jute geotextile performed satisfactorily in controlling soil erosion and helped in growth of vegetation. Bitumen treated jute geotextile has been used on the bank slope of Nayachar Island, in the river Hooghly, West Bengal, India for erosion control in 1992.[52] The undisturbed bank after 11 years implies that jute geotextile performed its designed functions and helped in natural consolidation of the bank soil. Application of treated woven jute geotextiles along with appropriate engineering measures was also done for prevention of riverbank erosion in the river Ichamati, West Bengal, India. Jute accounts for less than 1% of total geotextile use, despite the technical advantages and low cost of it.

2.6.1 Characteristics of jute geotextiles

Jute geotextiles have high potential application in the field of different civil engineering constructions. Woven from heavy and coarse jute yarn and having wide open mesh structure, jute geotextile is the ideal erosion control material for soil slopes under all climatic conditions.[53] Made from a natural fibre, jute geotextile is eco-friendly, biodegradable and decomposing and thereby it adds to the soil rich organic nutrients.[54] Being free from toxins and plasticizers it has no pollutants to run-off into ground water or to disturb the ecological system.[55] Its unique mesh construction leaves plenty of rooms for plants to grow and light to enter between the strands and its natural water absorbing capacity helps conserve soil moisture and anchor soil firmly. During water-flow each strand of jute geotextile forms a mini-dam that traps seeds and soil particles and reduces run-off velocity creating a microclimate conducive to germination of seeds and growth of vegetation to conserve soil.[56] Weighing 500 gsm or more it will not be easily lifted by wind or the flowing water. It is flexible enough to follow any type of surface contour (drapability). Any variety of grass or ground cover can be selected to fit site and climatic condition for use of this soil saver. Jute geotextile can be used in conjunction with all standard construction and building techniques. Huge quantity of these products is very easy to produce by jute mills on bulk scale and can be tailor-made in designing JGT for different purpose. Civil engineering works would gain by improved performance and/or by reduced costs.

2.6.2 Property advantages and specific end-uses of jute geotextiles

Jute geotextiles have high strength and modulus, good dimensional stability and ability to withstand initial stresses of road construction, heaviness and appreciable thickness, good draping quality, stiff body preventing differential settlement on soil, high permittivity and transmitivity, irregular surface morphology preventing lateral and rotational slides, high water absorption performing well in filtration and drainage and soil consolidation (caking) functions, soil friendliness and addition of nutrients to the soil after degradation, eco-compatibility, vegetation support, easy availability, low cost and agro-renew ability. Jute geotextile finds its application in surface soil erosion control in slopes and plains, stability of embankments, strengthening of sub-grade soils in roads, protection of banks of rivers and waterways, sub-surface drainage, soft soil consolidation, etc.[57-59] Thus, jute geotextile withstands stresses in the constructional phases, prevents intermixing of different soil layers, acting as separator, performs filtration function and also controls lateral dispersion, subsidence and slides.

2.6.3 Potential areas of application of jute geotextiles

Jute geotextiles are used in various application areas related to civil engineering construction.[60, 61] This natural geotextile is used as impermeable sheet of synthetic coated non-woven fabric to prevent seepage, behind structural defense formed of concrete slabs, stores, cement mortar, etc. Needle punched jute non-woven geotextile is used as permeable filter cloth to permit seepage but prevent loss of soil, behind structural defence formed of concrete slabs, stone, gabions, etc. Jute geotextiles combined with other fabrics are used as reinforcement fabrics to protect vegetation to strengthen soil and blocks.[62] Jute geotextile protects slopes in earthen embankments, hill slopes, dumps and heaps of granular materials like fly ash in thermal power plants.[63] This natural geotextile stabilizes embankments and controls erosion in banks of rivers, waterways, canals, etc. Construction of rural road and pavements and management of subsidence of railway tracks have also witnessed the tremendous potential application of jute geotextile. Construction of concealed drains especially in hill roads and consolidation of any type of soft soil by pre-fabricated vertical jute drains are the innovative applications of jute geotextile.[64] Management of watersheds and prevention of denudation of arid and semi-arid lands are some of the other promising areas of application of jute geotextiles. Jute geotextiles are now being widely used in greater way in the field of agricultural, horticulture and forestry sectors as sunscreens, plant nets, wind shield, harvesting nets, field-nets for protecting crop from birds, weed management and over all protection, mulching on seed bed, soil conservation, development of forests in semi-arid zones, nursery pots and nets, nursery shed, etc. Jute geotextile holds an edge over its man-made counterpart in geo-environmental applications because of its bio-degradability.[65]

2.6.4 Types of jute geotextile

There are three basic types of jute geotextiles based on constructional features and these are woven jute geotextile, open weave jute geotextile commonly known as soil saver and non-woven jute geotextile.[66] The process of manufacture of each is different as also their end-uses. All geotextiles are manufactured for specific end-use and in consideration of soil features necessitating refinement of fabric specification especially in regard to tensile strength and porometry. Woven and non-woven jute geotextiles are similar to their man-made counterparts.[67] There is however a difference in physical features in so far as open weave jute geotextile is concerned. Open weave jute geotextile possesses three-dimensional features unlike open mesh synthetic geotextiles. The intrinsic properties of jute impart high tensile strength, high tenacity, high moisture absorbing ability, high stiffness, low elongation, excellent drapability and high flexibility.[68]

2.6.4.1 Woven jute geotextile

This first type of jute geotextile has established its effectiveness in performing functions of separation, filtration, drainage and initial reinforcement. Woven jute geotextile is a fabric, which is manufactured by interlacement of warp and weft jute yarns set at right angle to each other through a machine called loom. The terms *warp* and *weft* are used to distinguish between the two different directions of yarns as shown in Fig. 2.8. *Warp* defines the longitudinal yarn, i.e. the direction in which production proceeds (also called *machine direction or MD*). *Weft* defines the transverse direction, i.e. running width-wise of the fabric (also known as *cross machine direction or CD*). The jute industry usually produces different broad categories of woven jute geotextile having fabric weights ranging from 600 gsm to 1200 gsm as per the end-use requirements.

2.6.4.2 Open weave jute geotextile

Open weave jute geotextile has an open structure similar to a net integrally connected two sets of parallel yarns which are interlaced at right angles to each other with pore size of about 20 mm × 25 mm. Figure 2.10 shows a typical open weave jute geotextile.

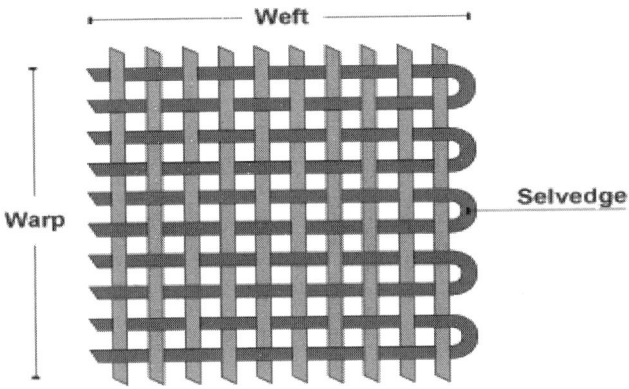

Figure 2.8   Pattern of weft and warp yarns in a woven jute geotextile.

Figure 2.9   View of a typical woven jute geotextile.

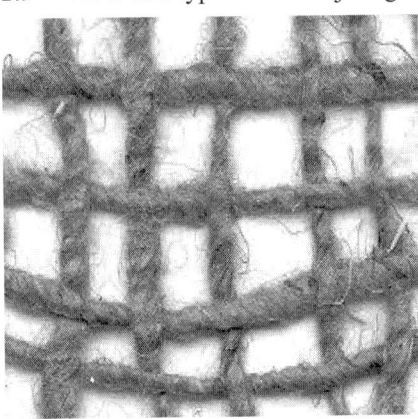

Figure 2.10   View of a typical open weave jute geotextile (soil saver).

This category of jute geotextile has been in use for years for arresting of surface soil erosion, slope protection and embankment stabilization. The jute industry usually produces three broad categories of open weave jute geotextile featuring weight per unit area which are 200 gsm, 500 gsm and 750 gsm mainly for control of soil erosion.

2.6.4.3 Non-woven jute geotextile

Non-woven jute geotextiles are obtained by processes other than weaving which include continuous laying of fibres on a moving conveyor belt bonded by mechanical such as needle punching, thermal and chemical processes. This category of jute geotextile is low in tensile strength and is principally applied for facilitating drainage and sometimes used in combination with the woven jute geotextile to exploit the dominant physical properties of the two types. In relation to synthetic fibre consumption in the field of nonwovens, the role of natural fibre is only to the extent of 2%. However, there is a wide scope for its exploitation because of its low price, ease of availability and biodegradability. It is also suitable for soil erosion control as geotextile material. The reason for selection of jute fibres for nonwovens are performance, safety, natural image, stability, mechanical aspects such as wet tensile strength, degradability, absorption and lastly low priced and easy availability and renewable source jute being the annual crop. A typical non-woven jute geotextile is shown in Fig. 2.11.

Figure 2.11 View of a typical non-woven jute geotextile.

In civil engineering application, non-woven jute geotextiles function as stabilizing, reinforcing, separating and filter media and are used in construction of temporary roads, highways, drainage, trenches. Agricultural and horticultural applications of jute non-woven are of recent developments. The most successful non-woven product is crop cover protecting seedling and young crops against the natural elements, insects, birds, etc. Among the various methods of manufacturing non-woven jute geotextiles, needle punching is the most versatile method and is widely used in jute and textile industry. Significant developments have taken place in recent years in the product design and their production technology for producing high quality needle punched synthetic non-woven fabrics, particularly suitable for use in the field of geotextiles, filter fabrics, agro textiles and automotive textiles. However, considering ecological aspects, disposability problem of synthetic non-woven jute geotextiles is a great concern. Due to this reason, jute, as a natural, biodegradable fibre, can be considered as a potential candidate for designing and manufacturing different non-woven products for certain end uses. The above stated technical textile products require some special mechanical and functional properties, which can be achieved through proper selection of suitable quality of raw material, suitable process parameters, and product dimensions like fabric weight/unit area, bulk density, thickness, etc. and machine parameters like type of needle and needle parameters, punch density and depth of needle penetration, etc.

2.6.5 Market potential of jute geotextiles

The erstwhile golden fabric jute is now reduced in common perception to mere 'gunny bags'. While the use of jute in packaging, home décor, etc. is well known, the use of jute as geotextiles is a largely unexplored area, although it can offer vast benefits to the indigenous industry and agro-economy overall as shown in the pie diagram in Fig. 2.12. The traditional market of jute is squeezing as a result of entry of synthetic yarns made of polyamides, polypropylene and other derivatives of petroleum chemicals in the market. Therefore, the jute industry can only be revived with its diversified uses. Out of all the probable uses of jute, the use of jute geotextile can be one of the potential areas to revive jute industry as a whole. Therefore, the increased offtake of jute will help in poverty alleviation in jute-growing areas and in improving the living conditions of farmers and workers. So its versatility is only coming to light now as the world looks on for this natural golden fibre to take over with the ideal solution for the modern civil engineering constructions like roads, railways, river embankments, soil erosion control, agro mulching, etc. Be it in conserving the soil and the environment as well as its applications like civil, environment engineering, etc. which are essential for the progress of civilization mainly in the areas of road construction. Geotextiles are being widely used in developed countries like United States of America (USA), United Kingdom (UK), Switzerland, Italy as well as other foreign countries and to some extent underdeveloped countries also. Therefore, jute geotextile should have a share of total geotextile market.

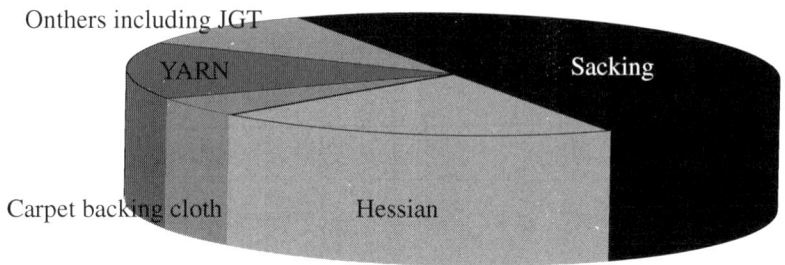

Figure 2.12 World Jute Products.

Source: *http://www.ruraldevelopment.info/Pages/JuteandHardFibres.aspx*

2.6.6 Present status of jute geotextiles

Jute geotextiles have been popularly used for protecting slopes and riverbanks and in slope stabilization applications for several decades even before geosynthetics came into the scene. It is only now that the real value and potentiality of natural geotextile is being assessed and benefits realized in a number of ways.[69] In India, geosynthetic applications started with the use of non-woven and woven textiles from 1984 onwards.[70] Limited trials were made in specific application segments like canal lining, track stabilization and road construction, etc.[71] Till 1994, in India only non-woven, woven and extruded geonets were available. But during 1994–95, the new generation of geosynthetic products arrived in Indian subcontinent. Geosynthetics have since been used extensively in many more projects like waste disposal and water resources projects, power projects, airports and hilly areas. Besides geosynthetic materials, natural materials like jute and coir geotextiles are also available in India in abundance, which have also been utilized for different applications.[72–79] Prefabricated vertical jute drain developed by Ramaswamy is quite well known by the name of fibre drain. The fibre drain is popularly used in Indonesia for soft deep-seated clay consolidation purposes. Being 100% biodegradable, it is the only environment-friendly drain available today, which makes it acceptable even by the developed countries. The global market of geotextiles is very fast expanding.[80] In the United States, the European Union, Japan along with some subcontinent of Southeast Asia which are the largest markets for geotextiles, there are major concerns as to environmental degradation and poor water quality caused by uncontrolled soil erosion. In order of importance, these markets are North America, which currently consumes around 50% of global rolled erosion control products, Western Europe, which consumes around 38%, and the rest of the world, which consumes the remaining 12% of production. Of course the regions like North America and Western Europe together account for some 88%

of world consumption and are therefore the prime target markets. Over the next few years these markets are forecasted to expand at around 5% per annum although within Western Europe there are countries where general geotextile consumption is expanding more rapidly. The current share of man-made geotextiles is 95% of the total global consumption vis-à-vis a meager 1% share of jute geotextile. A recent study by Tata Economic Consultancy Services (TECS) reveals that the global geotextile industry is expected to expand @ 10% to 15% per annum.[81] Though the projected rate of growth sounds rather optimistic, there is no denying the fact that the global geotextile market has significant growth potential. According to the said study by TECS, the global demand of geotextile may reach a figure of 2475 million sq. metres at the end of the current fiscal considering 10% annual growth rate taking 2000–01 as the base year. If 4% of the global demand of geotextile can be clinched by jute geotextile, the figure works out to 99 million sq. metres or around 49,500 metric ton (taking average fabric weight of jute geotextile as 500 gsm). The domestic market of jute geotextile is also showing up, though not consistently. According to reliable sources, the consumption of jute geotextile reached a figure of around 2,500,000 sq. metres (around 1250 metric ton) in 2005–06 compared to 393,000 sq. metres in 2001–02 (around 196.5 metric ton) which is a jump of more than 6 times.[82, 83] This is a sure sign of increasing acceptability of the product in the country, but JGT has still a long way to go to reach the target. Way back in 1991, International Trade Centre UNCTAD/GATT made an assessment of jute geotextile market in Europe. There exists a good demand of jute geotextile in West Europe especially in France, Germany, Italy, Switzerland, UK, Holland and Belgium, however favour man-made geotextiles. USA is the biggest importer of jute geotextile under the brand name of 'Soil Saver'. Jute geotextile is at present exported to 20 other countries from India, where soil saver is dominating maximum share. But its potential in Japan, Korea, Australia and Middle East countries still remains to be tapped. The market for geotextiles is projected to reach $8632.83 million by 2019, growing with a CAGR of 10.59% between 2014 and 2019. Asia-Pacific dominated global geotextiles market in 2013, in terms of value. The global geotextiles market is growing owing to the government policies and environmental norms supporting the commercial usage. Globally, governments in developed countries are making significant investments to secure long-term future and their attitude towards socio-economic benefits is supporting the growth of geotextiles market. In India, the government aims to invest $1 trillion in infrastructure during their Twelfth Five-year plan (2012–2017). Presently, the global demand of jute geotextile is 3 million tons.

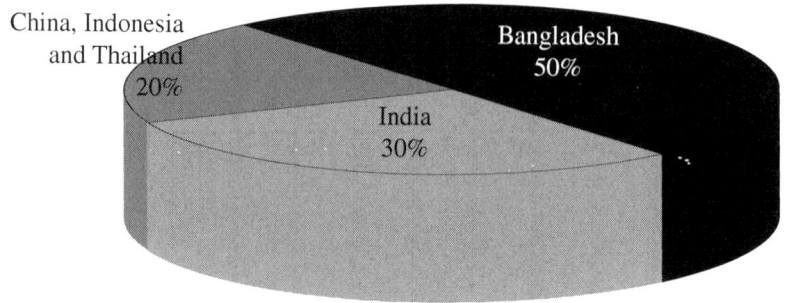

Figure 2.13   Production of jute geotextile in China and other Southeast Asian countries.

Source: International Jute Study Group (IJSG), http://www.jute.org /wjp/ cp_indonesia.htm

*Courtesy to: *Ranganathan, S.R., Development and Potential of Jute Geotextiles and Geomembranes, Vol. 13, No. 67, pp. 221–234, Elsevier Applied Sciences, Oxford, 1994.*

***Sanyal, T. and Chakravorty, K., Susceptibility of geojute to estuarine environment, International Seminar on Biocomposits, pp. 59–66, Delhi, 1994.*

References

1.   Maiti, R.K., A study of the microscopic structure of the fibre strands of common Indian bast fibres. Economic Botany 33(1): 78–87, 1979.

2.   Atkinson, R.R., Jute Fibre to Yarn, Chemical Publishing Co, New York, USA, pp. 13, 27, 36, 1965.

3.   Sanyal. T., Geosynthetics with Natural Fibres, In: Proc. a Seminar—Workshop Geosynthetics India, Indian Institute of Technology, New Delhi, India, pp. 362–367, 2004.

4.   Kundu, B.C., Basak, K.C. and Sarkar, P.B., Jute in India, The Indian Central Jute Committee, Kolkata, India, pp. 54–55, 1959.

5.   Jute, Kenaf and Allied Fibres, FAO Statistics, Food and Agriculture Organisation of the United Nations, Rome, June, 2007.

6.   Indian Jute, A Bulletin of Jute Manufactures Development Council, Kolkata, XIX(1), 2, June, 2007.

7.   Basu, N.C., Raw jute production in India—Problems and prospects. In: Paper presented at National Workshop held at the Central Research Institute for Jute and Allied Fibres (CRIJAF), Barrackpore, West Bengal, India, 1997.

8. Singhvi, G.M., Indian Jute a new symphony. In: Proc. Int. Jute Symposium on Indian Jute Sector and It's Relevance in 21st Century, Kolkata, India, pp. 22–28, 2003.

9. Data Book on Jute, first edition, Mitra, B.C., (Ed.), National Institute of Research on Jute and Allied Fibre Technology, Kolkata, India, 1999.

10. Atkinson, R.R., Jute Fibre to Yarn, Chemical Publishing Co, New York, USA, pp. 42, 1965.

11. Kundu, B.C., Basak, K.C. and Sarkar, P.B., Jute in India (The Indian Central Jute Committee), Kolkata, India, pp. 108, 1959.

12. A Treatise on Physical and Chemical Properties of Jute, International Jute Council (International Jute Organisation), Dhaka, Bangladesh, IJC (X)/11, pp. 20, 24, 30, August, 1988.

13. Barkar, S.G., Journal of the Textile Institute, 30, 273, 1939.

14. Chattopadhyay, D.P., Colourage, 45(5), 23, 1998.

15. 50 years of Research 1939–89, Pandey, S.N. and Anantha Krishnan, S.R. (Eds.), Jute Technological Research Laboratories, Indian Council of Agricultural Research, Kolkata, India, pp. 44, 55, 62, 63, 1990.

16. A Treatise on Physical and Chemical Properties of Jute, International Jute Council (International Jute Organisation), Dhaka, Bangladesh, IJC (X)/11, pp. 50, August, 1988.

17. Kaswell, E.R., Textile Fibres, Yarns and Fabrics, Reinhoid Publishing Corporation, New York, USA, pp. 11, 112, 1953.

18. Goswami, B.C., Martindale, I.G. and Scardino, F.L., Textile Yarns: Technology, Structure and Applications, John Wiley and Sons, New York, USA, pp. 17, 26, 1977.

19. Meredith, R., Journal of the Textile Institute, 36, 107, 1945.

20. Carlene, P.W., Journal of the Society of Dyers and Colourists, 60, 232, 1944.

21. Barkar, S.G., The Moisture Relationships of Jute (IJMA), Kolkata, India, pp. 122, 1939.

22. Preston, J.M. and Nimkar, M.V., Journal of the Textile Institute, 40, pp. 674, 1949.

23. Preston, J.M., Modern Textile Microscopy, Emmot and Co. Ltd., London, UK, pp. 40, 1933.

24. Bandyopadhyay, S.B., Textile Research Journal, 21(9) 659, 1951.

25. Chatterjee, H., Fibres (Natural and Synthetics), pp. 343, 1954.

26. Mohanty A.K., Misra M. and Drzal L.J., Sustainable bio-composites from renewable resource. Roy, A., Jute Chronicle, 3, 134, 1968.

27. Opportunities and challenges in the green material world. Journal of Polymer & Environment, 2002.

28. A Treatise on Physical and Chemical Properties of Jute, International Jute Council (International Jute Organisation), Dhaka, Bangladesh, IJC (X)/11, pp. 20, 24, 30, August, 1988.

29. Guha-Roy, T.K., Mukhopadhyay, A.K. and Mukherjee, A.K., Textile Research Journal, 54, pp. 874, 1984.

30. Macmillan, W.G., Indian Textile Journal, 67, pp. 338, 1957.

31. Sadov, F., Karchagin, M. and Matetsky, A., Chemical Technology of Fibrous Materials, MIR Publisher, Moscow, Russia, pp. 22, 1978.

32. Indian Standard: 271-1987, Bureau of Indian Standard, New Delhi, 1987.

33. Sanyal T., Khastagir A.K., Preponderance of Jute as Geotextiles. Geosynthetics Asia, Bangkok, Thailand, December, 2012.

34. Stout, H.P., Handbook of Fibre Science and Technology, Lewin, M. and Pearce, E. M. (Eds.), Marcel Dekker Inc., New York, USA, Vol. IV, pp. 701, 1985.

35. Ghosh, P., Samanta, A.K. and Das, D., Indian Journal of Fibre Textile Research, 19, 277, 1994.

36. Ghosh, T., Handbook on Jute, Plant Production and Protection, Paper No. 51, FAO, Rome, 1983.

37. Hoque, M.Z., An Introduction to Jute/Allied Fibres Properties and Processing, International Jute Organization (IJSG), 1992.

38. Booth, J.E., Principles of Textile Testing, CBS Publishers & Distributors in association with Butterworth Heinemann Ltd., pp. 510, 1968.

39. Martindle, J.G., Journal of the Textile Institute, 41, pp. 340, 1950.

40. Townsend, M.W., Journal of the Textile Institute, 42, pp. 12, 1951.

41. Onions, W.J., Pickering, J. and Stables, W., Journal of the Textile Institute, 41, pp. 480, 1950.

42. Booth, J.E., Principles of Textile Testing, CBS Publishers & Distributors in association with Butterworth Heinemann Ltd., pp. 295, 296, 1968.

43. Venkatappa Rao, G., Banerjee, P.K., Shahu, J.T., Ramana, G.V., Geosynthetics—New Horizons, Asian Books Pvt. Ltd., New Delhi, pp. 1–8, 1994.

44. Ramaswamy, S.D. and Aziz, M.A., Jute geotextiles for roads. In: Proc. Int. Workshop on Geotextiles, Bangalore, India, 22–29th November, pp. 259–266, 1989.

45. Giroud, J.P and Perfetti, J., Use of fabrics in geotechnics. In: Proc. Int. Conf., Paris, France, Vol. 2, pp. 345–352, 1977.

46. Bhandari, P.K. and Garg, K.G., Geosynthetics in landslide control—A case record, Use of G Horrocks, A. R., The Durability of Geotextiles, Eurotex, Bolton, UK, pp. 1–3, 1992.

47. Geosynthetic in India Experiences & Potential, CBIP, New Delhi, pp. 335–362, 1989.

48. Banerjee, P.K., Development of New Geosynthetics Products through Blends of Natural Fibres, Environmental Geotechnology with Geosynthetics, Asian Society for Environmental Geotechnology, New Delhi, India, pp. 337–345, 1996.

49. Liekweg, M., Covering the globe in geotextiles, International Fiber Journal, 19(2), 10–22, April, 2004.

50. Rankilor, Peter R., Textile in Civil Engineering. Part I—geotextile. In: Handbook of Technical Textile, Horrocks, A.R. and Anand, S.C. (Eds.), pp. 357, 359–360, 2000.

51. Rankilor, Peter R., Membranes in Ground Engineering, John Wiley & Sons, Chichester, UK, 1980.

52. Sanyal. T., Control of bank erosion naturally—A pilot project in Nayachara island in the river Hooghly. In: Proc. National Workshop on Role of Geosynthetics in Water Resources Projects, New Delhi, India, 20–24th January, 1993.

53. Zanten, R.V.V., Geotextiles and Geomembranes in Civil Engineering, A.A. Balkema, Rotterdam, Boston, The Netherlands, pp. 6–7, 47–57, 1986.

54. Sanyal, T., Remedial Concept with Jute Geotextile in a Complex River Bank Erosion Problem. In: Proc. ICGGE, held at IIT, Mumbai, 8–10th December, 2004.

55. Choudhury, P.K., Das, A. and Sanyal, T., Jute geotextile and its application in civil engineering, agri-horticulture and forestry. In: Proc. the 5th Int. Symposium on Earth Reinforcement, Fukuoka, Japan, 14–16th November, pp. 239–242, 2007.

56. Sanyal. T., Proc. Int. Jute Symposium on Environmental Applications of Jute Geotextiles, Kolkata, West Bengal, India, pp. 143–160, 2003.

57. Siddique, Q. Islam, Proc. International Workshop on Jute Geotextiles: Technical Potential and Commercial Prospect, Kolkata, India, pp. 185, 2008.

58. Choudhury, P.K., Das, A. and Sanyal, T. Proc. 5th International Symposium on Earth Reinforcement, Fukuoka, Japan, pp. 239, 2007.

59. Mondal, J.N. Man-made Text in India, 32, pp. 151, 1988.

60. Ghosh, S.K. and Dutta, M., Geosynthetics—Its functional properties and potential applications, IE (I) Journal-TX, 87, 8–9, August, 2006.

61. Siddique, Q. Islam, Potential of Jute Geotextiles, Its Application and need for Standards and Regulation. In: Proc. Int. Workshop on Jute Geotextiles: Technical Potential & Commercial Prospects, Kolkata, 5–6th April, pp. 185–209, 2008.

62. Ramaswamy, S.D. and Aziz, M.A., Jute Fabric in Road Construction. In: Proc. 2nd Int. Workshop on Geotextiles, Las Vegas, pp. 359–363, 1982.

63. Ramaswamy, S.D. and Aziz, M.A., Proc. Int. Workshop on Geotextiles, Jute Geotextiles for Roads, Bangalore, India, 22–29th November pp. 1–3, 1989.

64. Sanyal, T. and Choudhury, P.K., Use of Jute Geotextiles in Road, River and Slope Stabilization. In: Proc. Indian Geotechnical Conference, Roorkee, India, December, pp. 1–7, 2013.

65. Ray, P., Samanta, A.K. and Datta, M., Virtue of Jute Fibre in the Global Technical Textile market. In: Int. Symposium cum Exhibition on Jute & Geo-Textiles, Czech Republic, 27th May, pp. 1–14, 2004.

66. Zanten, R.V.V., Geotextiles and Geomembranes in Civil Engineering, A. A. Balkema, Rotterdam, Boston, The Netherlands, pp. 6–7, 47–57, 1986.

67. Horrocks, A.R., The Durability of Geotextiles, Eurotex, Bolton Institute of Higher Education and Degradation of Polymers in Geomembranes and Geotextiles, Hamidi, S.H., Amin, M.B. and Maadhah, A.G. (Eds.), Marcel Dekker, London & New York, pp. 433–505, 1992.

68. Ingold, T.S., The Geotextiles and Geomembranes Manual, first edition, Elsevier Science Publishers Ltd., Oxford, pp. 2–7, 199–231, 1994.

69. Ramaswamy, S.D., Jute geotextile for surface and subsurface drainage. In: Proc. Regional Seminar, Dhaka, December, 1997.

70. Mandal, J.N., Potential for use of natural fibers in geotextile engineering. In: Fourth International Conference on Geotextiles and Geomembranes, 28 May–1 June, The Hague, The Netherland, Vol. 1, pp. 835–838, 1990.

71. Talukdar, M.K., Rakshit, A.K. and Ghosh, S.K., Application of nonwoven Jute Fabric on Flexible Paved Road. In: Proc. Int. Seminar on Biocomposites, Delhi, pp. 105–111, 1994.

72. Ramaswamy, S.D., Development in natural geotextiles and application trends—Resume of—Paper. In: Int. Seminar on Biocomposites, Delhi, December, pp. 29–34, 1994.

73. Ranganathan, S.R., Development and potential of jute geotextiles, Geotextiles and Geomembranes, 13(67), 221–234, Elsevier Applied Sciences, Oxford, 1994.

74. Juyal, G.P. and Singh, Gurmeet, Geojute for rehabilitation of steep mine spoil areas. In: CSWCR & TI, Debra Dun, Uttarakhand, 1994.

75. Sanyal, T. and Chakravorty, K., Susceptibility of geojute to estuarine environment. In: International Seminar on Biocomposits, Delhi, pp. 59–66, 1994.

76. Jade, B.D., Desai, A.N. and Bal Subramanian, N., Some studies on physical and mechanical properties of jute-based nonwovens for geotextile applications. In: International Seminar on Biocomposites, Delhi, pp. 193–202, 1994.

77. Rao, G.V. and Balan, K., Design and development of a natural fibre strip drain. In: Int. Seminar on Bicomposites, Delhi, pp. 87–96, 1994.

78. Sivaramakrishnan, R., Design, development and marketing of jute coir geotextiles. In: Proc. Int. Workshop on Jute Geotextiles, Calcutta, August, 1997.

79. Subbarao, C., Proc. Seminar on Applications of Geosynthetic and in Highway Engineering, December, 1995.

80. Siddique, Q. Islam, Potential of jute geotextiles, its application and need for standards and regulation. In: Proc. Int. Workshop on Jute Geotextiles: Technical Potential & Commercial Prospects, 5–6th April, Kolkata, pp. 185–209, 2008.

81. Sanyal, T., Goswami, D.N. and Choudhury, P.K., PMGSY pilot project with jute geotextile—Its features and prospects. In: Proc. National Conf. & Exposition on Rural Roads, Vigyan Bhawan, New Delhi, 22–24th May, pp. 95–102, 2007.

82. Sanyal, T. and Choudhury, P.K., Jute Geotextile in controlling reflection cracks in roads—A case study. In: Proc. IRC-67 Annual Session, Panchkula, Haryana, 17th–21st November, pp. 31-34, 2006.

83. Sanyal, T., Jute geotextiles—Its commercial prospects, Quarterly Bulletin on Investment, Industry and Trade in West Bengal, 3(2), 33–35, 2004.

Development of grey jute paving fabric (GJPF)

3.1 Idea behind designing and engineering of grey jute paving fabric

Geotextiles have witnessed unparalleled growth worldwide in recent years in the field of different civil engineering constructions. With the growing environmental concern the global emphasis is towards the application of eco-concordant, renewable green products and this has inclined towards the natural fibre-made fabrics.[1-3] Jute geotextile is increasingly gaining ground over its synthetic non-biodegradable and toxic counterpart in different significant civil engineering constructions.[4-6] Out of these the development of geotextiles in the past three decades had provided the strategies for enhancing the overall performance of the paved roadways.[7] Hence, there is an urgent need to design and develop a fabric as overlay on existing pavements to stay technically and economically competitive in the global market. Such a fabric will not only prove techno-economically viable but will also reduce the carbon footprint generation to a large extent.[8] In the recent past, several varieties of geotextiles either in the form of woven or non-woven have since been developed for a number of end uses mainly as underlay in road construction for strengthening the pavement structure by increasing the soil compactness apart from the other end uses.[9, 10] Some fabrics are also being used as interlayer for the prevention of the reflective cracking in the road. All the geotextiles that are being used in the geotechnical engineering like road construction, soil erosion control, slope management, agro textiles, etc. are either woven or non-woven. Nowadays multilayer/sandwich fabrics are being extensively used as technical textiles which have great market potential. As per the need of the day, Grey Jute Paving Fabric (GJPF)—a combined fabric of jute woven and non-woven of different number of layers, coalescing the advantages of both the woven and non-woven jute components have been produced and optimized in respect to their geotechnical property parameters for potential application in the field of geotechnical engineering. In fact, the properties of the woven and non-woven fabrics are of different nature and have their own characteristic features. If woven and non-woven jute fabrics can be combined with suitable mechanism to produce multilayer/sandwich fabrics, then properties of both the woven and non-woven fabrics can be assembled in the combined fabric which can be used as jute geotextiles. The developed fabrics may be treated further with suitable chemical or bitumen for potential application in the field of geotechnical engineering and automotive engineering. Therefore, an attempt has been made in this study to engineer and develop multilayer/sandwich jute fabrics with the combination of different layers of woven and non-woven fabrics by needle-punching system in a commercial jute mill in the existing set up of the mill to incorporate both the properties of woven and non-woven fabrics in the combined multilayer/sandwich fabrics which is expected to perform better than the performance of single woven or non-woven fabric. Physical, mechanical and hydraulic property parameters of the developed fabrics have been measured for optimization of the geotechnical property parameters for those developed multilayer/sandwich jute fabrics for potential applications in the field of geotechnical engineering as well as in bioengineering fields.

3.2   Experimental methods

The entire experimental operation for preparation of different multilayer/ sandwich fabrics with the combination of woven and non-woven fabrics was carried out in a commercial jute mill in its existing setup for production of woven and non-woven fabrics. Production of such fabric samples along with selection of raw jute fibres and fibre mixing (batch composition) has been carried out under four stages: (a) choice of raw jute fibre mixing, i.e. batch composition, (b) preparation of woven fabric samples, (c) preparation of single layered combined woven and non-woven fabric samples, (d) preparation of multilayer/sandwich fabrics.

3.2.1 Choice of raw jute fibre mixing (batch composition)

Raw jute fibres of different grades and quality including assorted root cutting from long jute, jute caddies, etc. have been used as shown in Table 3.1 for preparation of different woven and non-woven needle-punched jute paving fabric samples. The criteria for selection of the raw materials are fibre strength in terms of bundle tenacity, fibre fineness, breaking extension, bulk density, colour and lustre and moisture regain percentage.

Based on experience and economic considerations and depending upon common fibre quality criteria/attributes of jute fibres, different fibre-mixing processes (batch composition) have been used for preparation of all the paving fabric samples. The batch composition and oil in water emulsion that have been used for preparation of woven and non-woven fabric samples are provided in Tables 3.2, 3.3, 3.4 and 3.5, respectively.

Table 3.1   Physical characteristics of jute fibres used in the study

Fibre type	Fibre strength in terms of bundle tenacity (cN/tex)	Fibre fineness in terms of linear density (tex)	Break-ing exten-sion (%)	Bulk densi-ty (g/cm³)	Colour and lustre	Mois-ture regain (%)
Jute (TD₅ grade)	32.81	2.36	1.9	0.6	Creamish yellow	12.90
Jute (TD₇ grade)	31.20	2.76	1.7	0.6	Creamish brown	12.90
Root cutting of long jute from Bangla-desh	33.61	3.11	1.4	0.7	Dark creamish brown	12.40

Table 3.2   Batch composition of super hessian warp batch jute woven fabrics

Sl. No.	Batch composition	In percentage (%)
1.	TD₅ long jute reed without root	70
2.	TD₅ root cutting from long jute	20
3.	Root cutting of long jute from Bangladesh	10

Table 3.3   Batch composition of sacking weft batch jute non-woven fabrics

Sl. No.	Batch composition	In percentage (%)
1.	TD₅ long jute reed without root	50
2.	TD₅ root cutting from long jute	20
3.	Root cutting of long jute from Bangladesh	30

Table 3.4   Batch composition of caddies and waste batch jute non-woven fabrics

Sl. No.	Batch composition	In percentage (%)
1.	Jute-caddies and jute mill line waste	100

Table 3.5 Batching emulsion composition (oil in water emulsion)

Sl. No.	Composition	Long jute	**Root cutting and mill line waste**
1.	JBO (mineral oil type jute batching oil)	140.00 l	150.00 l
2.	Nonidet-P_{40} (non-ionic emulsifier)	0.60 l	0.60 l
3.	Bran culture (a solution of pectinase enzyme)	35.00 l	30.00 l
4.	Water	1124.40 l	1119.40 l
Total volume in litre		1300.00 l	1300.00 l

3.2.2 Preparation of woven fabric samples

Ten number of plain woven fabrics with different fabric weights, expressed in gsm, was prepared in the conventional jute loom by varying yarn parameters. The flow chart described in Fig. 3.1 explains the preparation of the jute woven fabric samples. These developed fabrics were tested for their physical and mechanical property parameters. The test results of all the woven samples are given in Table 3.6. Considering, the mechanical properties of those developed samples, sample number 9 is selected for production of multilayer fabric as woven fabric to achieve the maximum tensile strength of the developed fabrics and the results have been tabulated in Table 3.7.

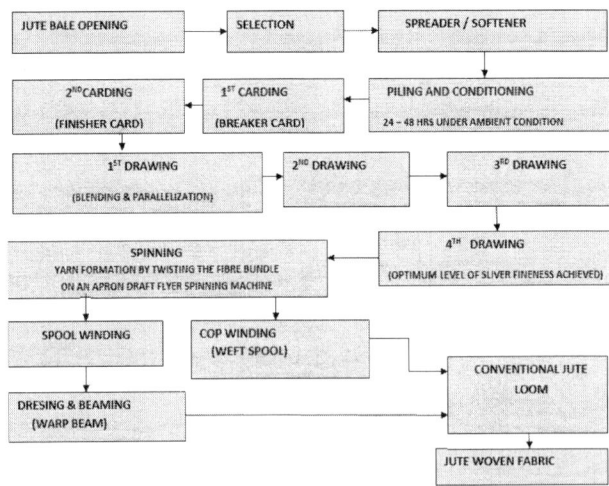

Figure 3.1   Flow chart describing the preparation of jute woven fabric.

Table 3.6   Test results of different produced woven fabric samples

Sample No.	Ends/dm × picks/ dm	Warp count × weft count (grist)	Areal density (g/m²)	Thick- ness (mm)	Wide width tensile strength MD × CD (kN/m)	Elonga- tion MD × CD (%)
1	24 × 24	7.0 × 7.0	122.00	0.85	5.80 × 4.40	3.0 × 5.0
2	25 × 24	8.0 × 8.0	140.00	1.13	4.43 × 4.52	5.0 × 6.0
3	25 × 23	8.2 × 8.3	145.00	1.22	4.58 × 4.88	5.0 × 7.0
4	33 × 26	7.5 × 8.0	165.20	0.77	7.38 × 4.00	5.0 × 7.0
5	37 × 24	8.0 × 7.5	183.00	0.88	9.17 × 4.90	4.0 × 7.0
6	37 × 30	8.0 × 7.0	193.00	0.93	6.75 × 6.25	6.0 × 6.0
7	37 × 32	6.0 × 8.0	195.00	1.06	8.68 × 8.20	4.0 × 6.0
8	38 × 33	6.0 × 8.0	200.10	1.08	8.88 × 8.62	5.0 × 5.0
9	38 × 34	7.0 × 9.5	205.00	1.03	10.07 × 11.43	5.0 × 5.0
10	38 × 33	6.5 × 9.5	210.00	1.12	7.99 × 8.38	5.0 × 6.0

Table 3.7   Test results of the selected woven jute fabric

Testing parameters	Values
1. Mass per unit area (gsm)	205.00
2.(a) Warp grist (lbs/spyndle)	6.90
(b) Weft grist (lbs/spyndle)	9.50
3. Thickness (mm)	1.03
4.(a) Ends/dm	38.00
(b) Picks/dm	34.00
5.(a) Wide–width tensile strength (kN/m), (MD × CD)	10.07 × 11.43
(b) Elongation-at-break (%) (MD × CD)	5.00

3.2.3 Preparation of non-woven fabric samples

The process sequences, followed for preparation of cross-laid jute needle-punched non-woven fabrics, starting from batching of raw jute fibres to final production of the non-woven fabric is shown by a flowchart diagram in Fig. 3.2. After preparation of properly fibre mixed rolls of jute from finisher card delivery, these finisher card rolls were taken to the feed of non-woven needle-punching machine line for web preparation and subsequent operations. Each and every stages of processing of jute fibres, starting from jute softener to the finisher card, proper quality control measures were adopted to maintain the desired level of quality of the jute non-woven needle-punched fabric. In needle-punching machine, prior to the needle loom, there is one carding unit having two pairs of worker and stripper roller. The delivery roller speed of this card was maintained at 29.00 m/min, the draft constant was 331.40, width of the card was 72 in., i.e. 182.88 cm, number of doubling used 10:1 and to get appropriate sliver weight/unit length, the range of draft pinion was used 15 T to 52 T, for all the prepared samples. Detailed machine parameters and process parameters setting of needle-punched non-woven plants including above stated carding unit are shown in Table 3.8. Thus, by keeping feed material constant with the above stated number of doubling using finisher card delivered mixed fibre roll (as per batch composition) different draft change pinions were used in the needle-punching line card to obtain a specific web weight per unit area (areal density) based on desired final areal density of the non-woven fabric without disturbing the draft constant, different setting parameters and delivery speed of this card. The fibre orientation of the delivered carded web was longitudinal after doffing from this card. Four numbers of such cross-laid non-woven fabrics of different areal densities (in the range of 200–300 gsm) were produced in a full scale commercial non-woven machine by M/S Fehrer Ltd., Austria (Model NL-9S). Lower gsm non-woven fabrics are not tried due to their poor tensile strength and higher gsm non-woven fabrics are also not tried as this will increase the thickness of the developed fabric apart from increment in the fabric weight. The web was therefore cross-laid with the help of a cross-lapper to obtain a non-woven web of layered structure with required web weight per unit area. During the cross laying, the running direction was changed to 900 and this changes the fibre orientation in the cross machine direction which is shown with the help of a schematic line diagram in Fig. 3.3. This cross lapper mechanism has a swinging depositing pattern, rendering a steep arm layering (camel back). The height of the steep arm-layering machine, i.e. the length of the lattice made aprons that swings back and forth, determines the maximum width of the web. The fibrous web coming out of the card was then fed to the feed lattice of the cross lapper which in turn feeds the web to the feed lattice of the needle loom in a cross laid fashion, the cross lapping angle used was 18° (to the perpendicular direction of feed). With the camel back, high quality of very uniform web can be produced to control precise mass per unit length. The speed ratio between camel cross lapper and cross-conveyor of this machine was adjusted to 18.2:1 and this ratio was maintained throughout the production of all jute non-woven needle-punched fabric samples.

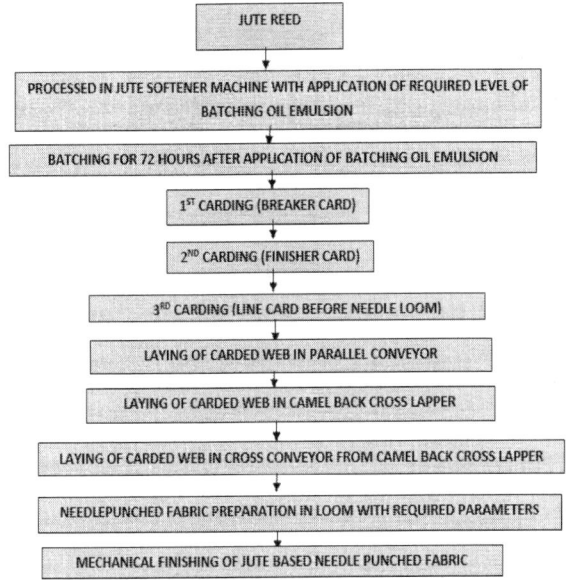

Figure 3.2   Flow chart describing the preparation of jute non-woven fabric.

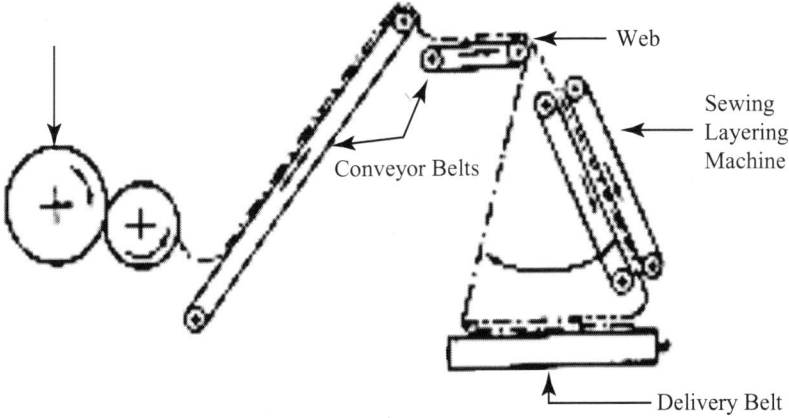

Figure 3.3   Schematic view of camel back steep arm Cross-Laying System.

Table 3.8   Detailed setting parameters of needle-punched non-woven production system including carding machine

Sl. No.	Parameters	Settings
1.	Carding machine (needle loom line card)	Roller and clearer type with 1 pair of worker and stripper roller
2.	Card doffer width (max.)	182.88 cm, (72 in.)

3.	Draft constant of card	331.40
4.	Card doffer speed	29.00 m/min
5.	Cross lapper type	Steep arm camel back type
6.	Cross lapper speed	31.10 m/min
7.	Feed lattice (of cross lapper) speed	1.71 m/min
8.	Cross lattice speed feeding to needle loom	2.36 m/min (during needling of feed web)
9.	Entry roller speed to the needle loom	2.36 m/min
10.	Delivery roller speed of needle loom	2.40 m/min
11.	Needle board strokes/min	450, 500, 570
12.	Number of needles/linear metre	3296
13.	Maximum needle loom working width	2.3 m
14.	Width of web fed to the needle loom (approximately)	2.05 m
15.	Width of needle-punched non-woven fabric produced from the needle loom	2.00 m
16.	The speed ratio between camel cross lapper and cross-conveyor	18.2:1
17.	Action of needle with respect to horizontal web movement	Needling from top at an angle 90° to horizontal web movement

3.2.3.1 The needle-punching machine

A full-scale commercial grade non-woven needling machine, Type–NL-9S, produced by M/s Fehrer Ltd., Austria has been used in this work. Specification of such needle-punching machine is given in Table 3.8. The feed lattice of the needle loom passes the multi-layered cross-laid web along with the woven fabric which is added to the feed lattice, to the needling zone for producing single layer combined woven and non-woven fabric. Therefore, the combined layer is fed on an endless belt to pass into the needle loom between the stripper plate and the bed plate, which contains holes of 1/4th in. diameter that matches the needle arrangement in the needle board. The stripper plate maintains and controls the web and the bed plate locates the lower layer of the fabric, and thus, in association with the needle movement, fixes the amount of needle penetration; the stripper plate prevents the web or individual fibres from being carried back with the needles. The cycle of operations includes two distinct and alternating actions in this needle loom: the forward movement of the web when the needles are out, and the penetration and withdrawal of the needles with the continuous movement of needled web by take-up roller. Then, the needled fabric passes from the plates to the take-up roller. This system is shown with a suitable schematic line diagram in Figs. 3.4(a) and (b), respectively.

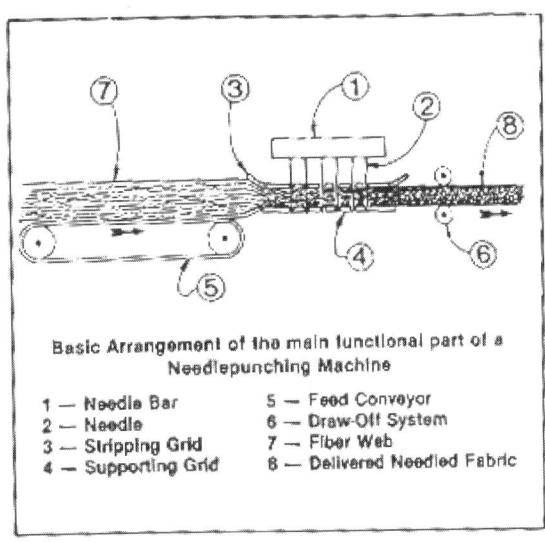

Basic Arrangement of the main functional part of a
Needlepunching Machine

1 — Needle Bar   5 — Feed Conveyor
2 — Needle   6 — Draw-Off System
3 — Stripping Grid   7 — Fiber Web
4 — Supporting Grid 8 — Delivered Needled Fabric

Figure 3.4   Schematic view of needle-punching process (American Hoechst Corporation).

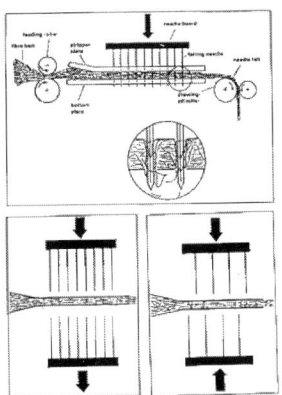

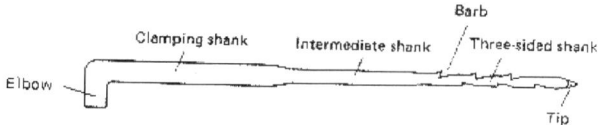

Figure 3.5   Schematic diagram of the needle.

The particular needle used in these experiments is shown in the Fig. 3.5 along with its detailed specification. Such felting needles are made of steel in a great variety of gauges, numbers and shapes of barbs and finishes. In this work, the diagonal pattern needle arrangement board was used. The board was 21 cm wide by 230 cm long, with a total of 103 needles/row width of the board, pitch of needle was 1.116 cm and distance between one row and the next successive row was 1.875 cm, which accommodates total 3296 needles only. For preparation of samples, only one particular type of needle board containing 25 gauge needles was used. Required needle depth penetration was adjusted as per the number of layers of woven and non-woven components of the sample through the external calibrated scale (in mm) with a pointer, where this system is directly connected with the needle board. Calibrated scale along with pointer assembly ranging from 0 to 40 mm has been fixed with one side of the loom.

The detailed specification of the needle is furnished in Table 3.9.

Table 3.9   Detailed specification of the needle

Sl. No.	Parameters	Settings
1.	Clamping shank	15 gauge
2.	Intermediate shank	18 gauge
3.	Three-sided shank	25 gauge
4.	Barb type	Regular
5.	Protrusion	Standard
6.	Needle length	31/5 in.
7.	Elbow length	1/5 in.
8.	Number of barb	3
9.	Diameter of shank before becoming three-sided	0.80 mm
10.	Depth of barb	0.12 mm
11.	Calculated cross sectional area of three-sided shank	0.5027 mm^2

3.2.3.2 Adjustment and calibration of needle-punching machine

The distance between the bed and stripper plates was adjusted so that the web will just pass through the gap without sticking or vibrating up and down. The gap is wider where the combined web and woven fabric enters and narrower where the consolidated combined fabric leaves. The needle penetration was adjusted by raising or lowering the bed of the machine, the needle stroke being kept constant; which was usually done while the machine was running. A scale indicates the position of the bed, and this is adjusted so that it reads zero when the needle tips are at the lowest point of the stroke and are just level with the top surface of the bed plate. Needle-penetration values quoted thus refer to the distance reached by the needle tips below the top of the bed plate, on which the lowest layer of the fabric rests. The feed-apron and take-up rollers are both driven from the crankshaft by eccentric and sprag clutches. The actual movement through the needling zone influences the amount of needling and that was controlled by the pull of the take-up rollers, by adjusting its speed during manufacture of different jute combined fabric samples. The punch density, i.e. number of punches on the surface of the feed in the web, is a complex factor and depends on the factors are density of needles in the needle board, rate of material feed, frequency of punching, effective width of the needle board and number of runs of the webs. Therefore, the punch density per unit area is a function of the number of needles per unit board width and the distance the web moves during each loom cycle. The following relation exists between the punch density per unit area and the number of needles per unit board width:

$$\text{Punch density / unit area } (cm^2) = \frac{\text{Needles / unit width (cm) of board}}{\text{Web movement / loom cycle (cm)}}$$

Where

$$\text{Web movement / loom cycle } (cm) = \frac{\text{Throughput speed (cm / min)}}{\text{Loom punches / min}}$$

Alternatively, if D is the needle density, namely, the number of needles per cm width of the board, and E is the length of fabric taken up per cycle of machine operation, the amount of needling M is defined as the number of needle penetrations per square cm of fabric, which is given by $M = D/E$. Thus, M can be determined, for a known needle density, from the calibrated take-up scale. If N is the number of machine cycles, or strokes per minute, the rate of fabric production will be EN or (DN/M) cm/min.

Table 3.10 Tensile strength of the samples with different folds

Sample No.	Tensile Strength (kN/m) and Elongation (%)							
	Single fold		Double fold		Triple fold		Four fold	
	Strength (warp × weft)	Elongation % (warp × weft)	Strength (warp × weft)	Elongation % (warp × weft)	Strength (warp × weft)	Elongation % (warp × weft)	Strength (warp × weft)	Elongation % (warp × weft)
1	5.8 × 4.4	3 × 5	11.35 × 9.82	4 × 5	18.25 × 16.08	4 × 5	25.12 × 18.54	5 × 5
2	4.43 × 4.52	5 × 6	9.15 × 8.60	5 × 5	15.65 × 16.08	4 × 5	19.85 × 20.15	5 × 5
3	7.38 × 4.00	5 × 7	15.50 × 7.90	4 × 6	21.10 × 14.50	4 × 5	21.78 × 21.37	5 × 5
4	9.17 × 4.90	4 × 7	16.85 × 10.15	5 × 6	27.55 × 14.45	6 × 5	36.55 × 26.18	6 × 7
5	6.75 × 6.25	6 × 6	15.80 × 12.05	5 × 6	18.45 × 17.12	6 × 5	27.12 × 24.38	5 × 6
6	10.50 × 11.50	5 × 5	19.82 × 21.67	5 × 5	29.48 × 33.45	5 × 6	38.42 × 42.12	5 × 6

Table 3.11 Test results of the different combined fabric (woven and non-woven fabric components)

Sample No.	Specification	Areal density (gsm)	Thickness (mm)	Tensile strength (kN/m) MD × CD	Elongation (%) MD × CD	CBR puncture (kN)	Bursting strength (kg/cm²)	AOS [O_{95}] (μm)	Permittivity (/s)
1	1 layer woven + 1 layer non-woven	398.8	3.80	10.44 × 8.05	5.0 × 6.0	0.91	15.48	410	5.41
2	2 layer woven + 1 layer non-woven	580.0	4.28	13.51 × 9.42	5.0 × 10.0	1.21	20.96	340	3.85
3	2 layer woven + 2 layer non-woven	850.0	5.20	17.60 × 9.20	5.0 × 14.0	1.89	28.82	250	2.33
4	2 layer woven + 3 layer non-woven	1120.0	5.80	19.90 × 17.25	5.0 × 15.0	2.12	30.15	190	2.01
5	3 layer woven + 2 layer non-woven	1035.0	6.80	25.00 × 26.00	6.0 × 15.0	2.86	36.00	170	1.69
6	3 layer woven + 3 layer non-woven	1275.0	7.25	23.85 × 28.35	6.0 × 16.0	3.25	41.52	180	1.35
67	3 layer woven + 4 layer non-woven	1408.0	10.29	28.95 × 31.25	5.0 × 16.0	3.76	64.86	135	1.25
88	4 layer woven + 3 layer non-woven	1480.0	8.21	29.15 × 31.50	5.0 × 16.0	3.95	77.52	140	1.31
99	4 layer woven + 4 layer non-woven	1506.0	8.47	31.48 × 32.07	6.0 × 17.0	3.98	66.28	110	0.96
110	3 layer woven + 4 layer non-woven	1520.0	8.40	30.85 × 32.35	6.0 × 17.0	4.28	59.60	120	1.17

3.2.4 Preparation of single layered combined woven and non-woven fabric samples

The tensile strength of the individual fabrics of different folds like double fold, triple fold and four fold have been tested. From the test result it is found that the strength of the different fold fabrics increases proportionally to the strength of the single fold fabric in both warp and weft way. The test results are supplied in Table 3.10. The increase in strength of the doubled fabric is found to be proportional due to the sharing of the developed load equally by the different layers as well as friction between the different layers during breaking of the layered fabric. So, to get the desired strength of the paving fabric in the above said range the samples have been planned to be prepared by combining different layers of woven and non-woven fabrics. Therefore, as per the working plan, ten numbers of cross-laid non-woven batts of different gsm (120–250 gsm) have been produced and subsequently combined with the selected woven fabric sample in the needle-punching loom for production of single layered combined fabric.

Figure 3.6   Feeding of the carded slivers for non-woven fabric preparation.

3.2.5 Preparation of the grey jute paving fabric sample

After testing and analysis of all the single layered combined fabrics they have been selected for production of the grey jute paving fabric samples with some desired specific property parameters as per end-use requirements. Initially, ten numbers of grey jute paving fabric samples were produced by suitable combination of single layered combined fabrics and woven fabric samples, as shown in Figs. 3.6–3.11, with the help of needle-punching machine in a commercial jute mill. The physical, mechanical and hydraulic properties of the different combined layers of woven and non-woven fabric components have been furnished in Table 3.11. The selection of a five layered combination of jute paving fabric is not just a mere arrangement or emergence by chance. In fact, prior to achieving the above mentioned combination, several other combinations of woven and non-woven jute fabric components have been tried to finally obtain that grey jute paving fabric which will be able to satisfy specifically those requirements that are needed to be deployed during the road construction.

Figure 3.7   Combination of the cross lapped batt and single layer woven fabric.

Figure 3.8   Combined layer of woven and non-woven fabrics.

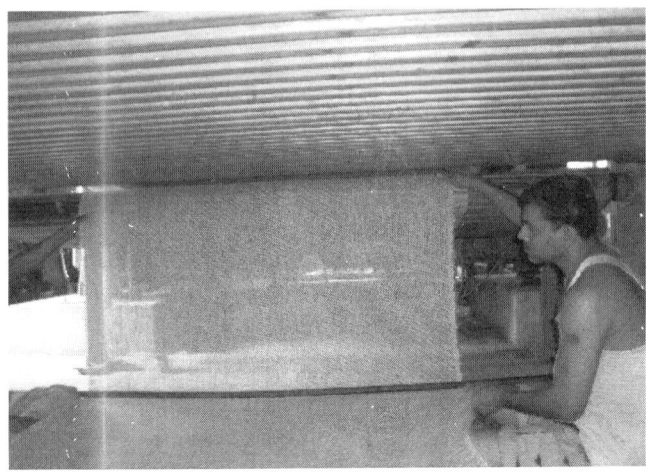

Figure 3.9   Combination of the combined layers of fabric with woven fabric layer.

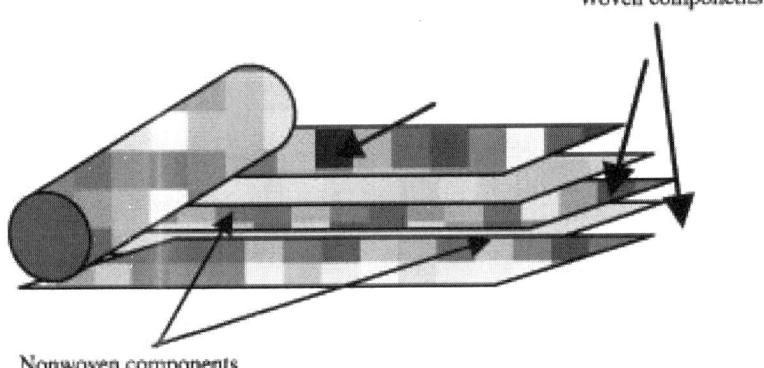

Figure 3.10   Schematic design of GJPF.

Figure 3.11   Finished roll of grey jute paving fabric.

3.2.6 Testing methods

In growing areas of specific uses of technical textiles, like filter fabric, geotextiles, etc. unified set of worldwide standards and test methods are currently not available, yet the activity towards such ultimate goal is very intense. Even in India, there is no such standard method of such testing standardized by BIS or any other standard testing agencies nationally. Moreover, the standard test methods that differ for general textiles and geotextiles materials are those involve, such as hydraulic, aerodynamic, soil compatibility, endurance and environmental properties; for which specific application oriented newer test methods have been designed separately considering the geotextile or other technical textile applications. However, for testing of geotextiles, some ASTM (American Society for Testing and Materials) and other standards have been emerged, even though these are not complete and exhaustive standards which are required for accessing different property parameters as per end-use requirement. But in the U.S., the ASTM has a standard committee specially organized for geotextiles material (D-35) testing methodology, which is very much useful and convenient for different geotextile applications in the field of civil engineering construction The safest and ideal way of selecting the fabric can be done through specific case study/end-use performance test of the products. But incidentally, very few historical case studies are available for jute and other geotextiles for comparison of their efficacy as well as potential application in the respective filed of geotechnical engineering. As a result, a detailed evaluation of the fabric criteria for a particular geotextile application is to be made on the basis of standard laboratory tests. On the basis of product and process conformity as well as depending upon different levels of produced fabric properties relevant for different geotextile application in the field of geotechnical engineering, those are broadly divided into four groups, namely: (a) physical properties, (b) mechanical properties, (c) hydraulic properties and (d) endurance and miscellaneous properties. Therefore, a series of standard test are performed on all the produced fabric samples to determine different geotechnical properties in the machine direction as well as cross machine direction following the ASTM standard as stated in Table 3.12.

Table 3.12 ASTM standard test methods followed to carry out the testing of the fabric samples

Sl. No.	Test parameters	ASTM standard
1.	Mass per unit area	D5261-92(2009)
2.	Fabric thickness	D5199-12
3.	Tensile properties of geotextiles by wide width strip method	D4595-11

4.	Static puncture resistance	ASTM D6241-04 (reapproved 2009)
5.	Bursting strength–diaphragm	IS:7016 (Part 6)-1984
6.	Permittivity	ASTM D4491-99a (reapproved 2014)
7.	Apparent opening size (AOS)	D4751-12

3.2.6.1 Conditioning of test fabric samples

The entire range of jute-based needle-punched non-woven paving fabric samples was conditioned according to an ASTM standard using standard temperature (21°C ± 2°C) and humidity (65% ± 5% R.H.) for 24 h before commencement of any testing work.

3.2.6.2 Selection of fabric samples for testing work

Small size samples used for normal textile testing cannot generally be regarded as appropriate for technical textile/geotextile testing. These small samples have only a limited usefulness in assessing the properties of a fabric relative to its engineering end use, samples-tested on modified or specially developed apparatus provide much more appropriate data. Therefore, test samples were selected in such a way that it could represent the whole population of the fabric and the piece of fabric cut out for the laboratory test should be at least 1 m long with full width of the fabric. No samples have been taken from nearer than 50 mm to the selvedge of the fabric samples.

3.2.7 Measurement of weight per unit area of the fabric samples

The specifications for mass per unit area for any geotextile fabric have direct influence on mechanical, hydraulic and other geotechnical properties. Weight/unit area of needle-punched non-woven jute fabrics (with or without bitumen treatment) was determined as per ASTM D 5261-92 standard testing method. For each sample, fifteen numbers of samples of each 250 mm × 250 mm dimension were cut from various locations over the full width of the fabrics and then weighed in precision electronic balance and then average weight per unit area was calculated for all those samples.

3.2.8 Measurement of fabric thickness of the fabric samples

Thickness is one of the basic physical properties used to control the quality of non-woven geotextiles. The normal thickness of any geotextile fabric is determined by observing the perpendicular distance that a moveable plane is displaced from a parallel surface by a specified pressure 2 kPa for geotextiles as recommended by ASTM D5199-12 standard testing method. The accuracy level for determination of fabric thickness is very essential as this data is prerequisite input for determining some important fabric parameters of geotextile fabrics, such as permeability of both of the air and water, permittivity, porosity, etc. The thickness of the fabrics was measured using Analogue Thickness Gauge (AIM-241), AIMIL Ltd., New Delhi, which is suitable for determining maximum thickness of 10 mm or 25 mm, provided with two dial gauges, one for the range of 10 mm (with accuracy of 0.002 mm) and the other for the range from 11 to 25 mm (with accuracy of 0.01 mm) under a pressure foot area of 31.66 cm^2 having standard pressure of 20 g/cm^2 for a compression period of 20 s. For accurately determining the thickness of the fabric, it is to be cut in such a fashion so that the material will extended beyond the edge of the pressure foot by 1.0 cm in all directions. Fifteen such observations for each sample were taken randomly from different parts of the sample and the average was calculated.

3.2.9 Determination of tensile properties of the fabric samples by wide width tensile testing method

Non-woven geotextile has the tendency of necking down behaviour and hence wide width tensile testing method is used in both the direction (in machine direction and as well in cross machine direction) of the fabrics. This test method is designed to reproduce plane strain conditions as closely as possible without the use of lateral restrains. INSTRON 5982 Microprocessor controlled CRE type (constant rate of extension) universal testing machine was used to determine the tensile strength, breaking elongation as per ASTM D4595-11 standard testing method. A relatively wide specimen is gripped across its entire width in the clamps of CRE type Tensile Testing Machine operated at a prescribed rate of extension 10%/min, applying a longitudinal force to the fabric specimen until the specimen ruptures. Tensile strength and elongation of the test specimen are obtained from the computer software. The distance between the clamps is adjusted at the start of the test at 100 mm. One clamp is supported by a movable grip which will allow the clamp to move in the plane of the fabric. The machine is set at a constant speed of 10 mm/min. The specimen is mounted centrally in the clamps. This is carried out by having the two lines drawn 100 mm apart across the width of the specimen positioned adjacent to the inside edges of the upper and lower jaw. The specimen length in the machine direction and cross-machine direction tests, respectively, are parallel to the direction of application of the force. The tensile testing machine is started and continued running the test to rupture. The readings are recorded in the machine software. The machine is stopped and reset to the initial gauge position to record and report the test results.

3.2.10 Determination of static puncture resistance of the fabric samples

This test indicates inherent resistance against penetrating or puncturing of geotextile fabrics by any object during use. This test was performed using INSTRON 5982 Microprocessor controlled CRE type universal testing machine following ASTM D6241-04 (reapproved 2009) standard testing method. A test specimen is clamped without tension between circular plates and secured in a tensile testing machine. A force is exerted against the centre of the unsupported portion of the test specimen by a steel plunger attached to the load indicator until rupture occurs. The maximum force is the value of puncture strength. The load range of the tensile testing machine is selected such that the rupture of the fabric occurs between 10% and 90% of full scale load. The test specimen is centred and secured between the holding plates ensuring that the test specimen extends to or beyond the outer edges of the clamping plates. The machine is set at a speed of 50 mm/min and switched on. The running of the machine is continued until the puncture rod completely ruptures the test specimen. The puncture strength is recorded by the computer which is interfaced with the testing machine.

3.2.11 Determination of bursting strength of the fabric samples

By this method the bursting strength of geotextiles is determined by applying hydrostatic pressure at a controlled rate through a rubber diaphragm. A test specimen is held between two annular clamps under sufficient pressure to minimize slippage. The upper clamping surface, which is in contact with the test specimen, has a continuous spiral groove and the lower clamping surface has a number of concentric grooves. A circular diaphragm of pure gum rubber is clamped between the lower clamping plate and a pressure cylinder so that before the diaphragm is stretched by pressure underneath it the centre of its upper surface is below the plane of the clamping surface. A test specimen of 90×90 mm size test sample is placed on the lower clamp with the area to be tested centrally located and the M/C is run until the test sample ruptures. The pressure required to rupture the sample is found from the pressure indicator.

3.2.12 Measurement of water permeability and permittivity of the fabric samples

Water flow rate, permeability and permittivity are also very important selection criteria for geotextiles used in civil engineering works for filtration and drainage as one of the purpose in railway, road, dam, drains and other related areas. The geotextile, therefore, has to satisfy certain requirements of water permeability according to the characteristics of soil and the type of end use application. A standard (BTRA made) water permeability tester was used for determining water permeability and other related properties perpendicular to the plane of fabric. This test also refers to measurement of permittivity of a geotextiles, which is the volumetric flow rate of water per unit cross-section area per unit head, under laminar flow conditions, in the normal direction through a geotextiles. Its test methods describe procedures for determining the water permeability by permittivity method of geotextiles using constant head or falling head test procedures, as per ASTM D4491-99a standard testing method.

3.2.13 Test procedure followed

The circular test specimens of each fabric samples of 5 cm diameter were conditioned by soaking them in water for 24 h prior to testing work. Then one specimen was mounted on the sample holder. Water was then backfilled through the discharge tube in such a way that no gap remains on the bottom portion of the column. The two halves of the column were then assembled with the sample holder in between. The backfilling of water was then continued for sometimes to ensure that no air is trapped below the sample holder. Inlet tap was then opened and water was allowed to flow into the apparatus until the water level reaches the overflow from outlet. The height of the discharge tank or input water flow rate was then adjusted to obtain 50 mm head of water on the fabric specimen. After steady flow rate of water was reached, the quantity of flow (Q) was recorded and indicated by the rotameter scale, which indicates flow rate in litres per min (LPM). The water temperatures are to be measured and recorded from time to time.

3.2.14 Determination of water permeability of the fabric samples

The water permeability of fabrics/geotextiles has been calculated from the following equations derived from Darcy's law: Water permeability is expressed as the rate of flow of water under a differential pressure through the material. The quantity of water allowed to pass through the specimen from a specific area is to be measured and then it is to be converted for 1 m^2 area of a fabric. Hence,

Water permeability is expressed in m/s; and $V = Q/A \cdot t$ (or flow velocity, l/m^2/s)

where, Q = quantity of flow, in litre (l)

A = cross-sectional area of test specimen in square metre (m^2)

t = time for flow, in second (s).

Permittivity of fabrics/geotextiles is the volumetric flow rate of water per unit cross-sectional area per unit head under laminar flow conditions, in the direction perpendicular to the fabric surface.

Ψ = permittivity, $(s^{-1}) = (Q \times R_t)/(\Delta h \times A \cdot t)$

where, Q = quantity of flow, in litre (l)

Δh = hydraulic head difference across the specimen, in metre (m)

R_t = temperature correction factor determined by $R_t = U_t/U_{20}$

where, U_t = water viscosity at test temperature, (millipoise)

U_{20} = water viscosity at 20°C, (millipoise)

Coefficient of permeability, K can be determined from the permittivity using the equation:

$$K = \Psi \cdot T_g$$

where, K = permeability coefficient, (m/s)

T_g = nominal thickness of the geotextiles, (m)

Ψ = permittivity, (s^{-1})

3.2.15 Constant head method

A single layer of the geotextiles specimen is to be subjected to a unidirectional flow of water normal to the plane under a range of constant water head.

Table 3.13   Testing conditions and different parameters for the measurement of water permeability and permittivity

• Permeability (flow rate)	$l/m^2/s$ (m/s)
• Permittivity	s^{-1} (flow rate per unit water head)
• Working diameter of specimen	50 mm
• Water head	50 mm
• Sample preparation	Specimens are cut diagonally from various portion of a sample and allowed to soak in water for 24 h
• Number of specimens	15

Temperature correction: Water temperature correction can be calculated from the following relation:

$$R_t = U_t/U_{20}$$

where,

U_t = water viscosity at test temperature, (millipoise)

U_{20} = water viscosity at 20°C, (millipoise)

Table 3.14 Relation between test temperature and corresponding temperature correction factor (R_t).

Test temperature, °C	Temperature correction (R_t)
19	1.0240
20	1.0050
21	0.9761
22	0.9531
23	0.9311
24	0.9096
25	0.8892

Permittivity can be calculated from the following relation:

$$\text{Permittivity} = \frac{\text{Litres} / \text{sq.m} / \text{sec}}{\text{Water head in mm}}$$

Fifteen observations for each sample were taken randomly from different parts of the non-woven sample and finally the average was calculated.

3.2.16 Measurement of apparent opening size of the fabric samples

The apparatus for measurement of effective opening size of geotextile fabric has been specially designed to suit geotextiles of various thicknesses and grades. To make the process of sieving easier and quicker, electrically operated mechanical Sieve shakers were used for dry sieving. This test method covers the determination of apparent opening size (AOS) for a geotextile by sieving glass beads through a geotextile fabric. Apparent opening size of non-woven fabric samples was measured using AIMIL Ltd., Dry Sieve Test Apparatus (AIM-242) following ASTM D4751-99a standard testing method. As per this standard, the pore size was calculated for all of the fabric samples in using the standard O_{95} sieve was expressed in μm. For the above said AOS test, ten test specimens were cut from each category of fabrics to fit appropriately to sieve pan. Each specimen was pre-weighted in electronic balance and was then immersed in distilled water for one hour at standard atmospheric conditions. Finally, the specimens and glass beads were dried in oven until no weight change was recorded. The sieve frame containing the geotextile fabric and loose glass beads on it were then shaken laterally for a time period of 10 min, so that the jarring motion will induce the beads to pass through the test specimen as much as possible. 50 g of such dry glass beads of known size (in μm) was placed on the geotextile fabric mounted on the sieve pan and the fabric was then vibrated for 10 min at an appropriate frequency with amplitude sufficient to cause the glass beads to move over the mounted fabric surface horizontally but not allowing them to move vertically. The total weights of the glass beads of definite size passed through the test fabric was recorded and the percentage retained of glass beads retained or entrapped was calculated after 10 min of vibrating of the sieve. The latter was deemed to be the pore size percentile against the mean diameter of any specific grade glass bead. The procedure was then repeated with next smaller grade of available glass bead. The above said test procedures were repeated using the glass beads of next larger size fraction. Thus, the same trial was repeated using succeeding larger bead size fractions until the weight of glass beads that passes through the specimen, is within 5% or less. The trials were repeated until percentage (%) passing of glass beads decreases from a value greater than 5% to a value less than or equal to 5%. Finally, the value of pore size (AOS) was determined from the graph drawn between glass bead size (along X axis) and percentage of glass bead passing through (along Y axis) with a level of 95% retention.

The specifications of the final grey jute paving fabric that has been optimized and selected from the different grey jute fabric samples have been furnished in Table 3.15.

Table 3.15 Specifications of the grey jute paving fabric (GJPF)

Sl. No.	Parameters	Values
1.	Fabric weight expressed in gsm	1035.00
2.	Thickness expressed in mm	6.80
3.	Wide-width tensile strength expressed in kN/m (machine direction × cross-machine direction)	25.00×26.00
4.	Elongation at break expressed in percentage (machine direction × cross-machine direction)	6.0×15.0
5.	Bursting strength expressed in kgf/cm^2	36.00

References

1. Data Book on Jute, first edition, Mitra B.C. (Ed.), National Institute of Research on Jute and Allied Fibre Technology, Kolkata, India, 1999.

2. Jute, Kenaf and Allied Fibres, FAO Statistics, Food and Agriculture Organisation of the United Nations, Rome, 2007.

3. Indian Jute, *A Bulletin of Jute Manufactures Development Council*, xix (1), 2, 2007.

4. Aziz, M.A. and Ramaswamy, S.D., Some Studies on Jute Geotextiles and their Applications, Geosynthetic World, Wiley Eastern Limited, pp. 337–345, 1996.

5. Liu, Aimin, Jute-An Environmentally Friendly Product; International Commodity Organization in Transition, United Nation Conference on Trade and Development, 2001.

6. Hoque, M.Z., Study on the agro-ecological conditions of jute and kenaf producing countries, International Jute Organization, IJSG, 1996.

7. Hoque, M.Z., Study on the agro-ecological conditions of jute and kenaf producing countries, International Jute Organization, IJSG, 1996.

8. Inagaki, H., Progress on Kenaf in Japan, Abstract, Third Annual Conference of American Kenaf Society, Corpus Christi, Texas, 2000.

9. Hoque, M.Z., An introduction to jute/allied fibres properties and processing, International Jute Organization, IJSG, 1992.

10. Final Report on Environmental Assessment of Jute agriculture – Phase-I, International Jute Organization, IJSG, 2000.

11. Banerjee, P.K., Development of new geosynthetics products through blends of natural fibres, environmental geotechnology with geosynthetics, Asian Society for Environmental Geotechnology, 337–345, 1996.

12. Ray, A. N., Studies on Jute Based Needle punched Nonwovens using Multifactor experimental Designing Technique, Ph.D. Thesis, pp. 1–3, 2003.

13. Venkatappa Rao, G. and Balan, K., Durability of jute fabrics, environmental geotechnology with geosynthetics, Proc. 7th Asian Textile Conference, Asian Society for Environmental Geotechnology, 348–357, 1996.

14. Ranganathan, S.R., Development and potential of jute geotextiles and geomembranes, Geotextiles and Geomembranes, 13(6–7), 421–434, 1997.

15. Sanyal, T., Proc. Int. Jute Symposium on Environmental Applications of Jute Geotextiles, India, pp. 143–160, 2003.

4

A comprehensive idea about bitumen

4.1 Introduction

Before moving further into the subsequent chapters narrating the developed Bituminized Jute Paving Fabric (BJPF), the author thinks it is wise to give the readers a comprehensive idea about bitumen and its properties in this chapter. Bitumen which is also known as asphalt or tar is a black, oily, viscous, flammable material and a naturally occurring organic by-product of decomposed organic materials, is well known worldwide for its use in roof waterproofing, bridge-deck waterproofing, basement tanking, flooring, paving and for some heavy duty road surfacing and bridges. For more than a century, paved roadways which include the carriageways and the shoulders have been constructed out of bitumen as the major ingredient. Commonly a paved road becomes a candidate for maintenance when its surface shows significant cracks and potholes. The construction of asphalt overlays of an optimum thickness is the most common way to renovate both flexible and rigid pavements. But the cracks under the overlay rapidly propagate through the new surface which is a major drawback of asphalt overlays. The generation of these cracks have caused riding discomfort for the users, infiltration of water and subsequent reduction of the bearing capacity of the sub-grade and forced to undergo changes in the basic design method along with construction technique of the paved roads. The development of geosynthetics in the past three decades has provided the strategies for enhancing the overall performance of the paved roadways. The use of geosynthetic as interlay and overlay, can enhance the life of the overlay via stress relief, preventing seepage of water through the pavement and reinforcing the cracked old pavement. Thorough research work has revealed that jute and hot bitumen have excellent thermal compatibility. Thus bitumen-soaked jute overlays may be used as riding surface of roads for resurfacing the distressed riding surfaces on flexible pavements. Therefore, like bitumen, geosynthetics also play a vital role in civil engineering constructions. Starting from a small path created naturally due to frequent walks, the roadways in India have come a long way. Today the Rajpath of Delhi and Formula One Track nest in the same neighbourhood, while the national dream of connecting Srinagar, Porbandar,

Mumbai, Kanyakumari, Kolkata and Guwahati on the same road-grid, is inching towards reality. From the 'thread that binds the nation together' the Indian Road Network has metamorphosed into the 'weave that binds the nation together'.[1] The Road Transport in India has emerged as the dominant mode of transport and is vital to the economic development, trade and integration which rely on the movement of the people and goods. Moreover, Road Transport is the only mode which performs both mobility and accessibility functions, whereas all other modes like rail transport, water transport, air transport, etc. serve mobility function only. It is heartening to note that the Government of India has launched two flagship road sector programme like the National Highway Development Project (NHDP) and the Pradhan Mantri Gram Sadak Yojna (PMGSY) by connecting few of the rural roads along with some district roads for the last few years so that the fruits of development reach all sections of the society. Commonly a paved road becomes a candidate for maintenance when its surface shows significant cracks and potholes.[2] The high traffic intensity in terms of commercial vehicles, overloading of trucks and significant variations in daily and seasonal temperature of the pavement have been responsible for early development of distress like rutting, cracking, bleeding, shoving and potholing of bituminous surfacing.[3] The construction of asphalt overlays is the most common way to renovate both flexible and rigid pavements.[4] A minimum thickness of the asphalt concrete overlay may be required to provide an additional support to a structurally deficient pavement.[5] But the cracks under the overlay rapidly propagate through the new surface which is a major drawback of asphalt overlays.[6] The generation of these cracks have not only caused riding discomfort for the users, infiltration of water and subsequent reduction of the bearing capacity of the sub-grade on one hand but also forced to undergo changes in the basic design method and construction technique of the paved roads on the other hand.[7] Studies have revealed that properties of bitumen and bituminous mixes can be improved/modified with the incorporation of textile materials.[8] At present there is a huge global demand for suitable technical textiles to be used in the civil engineering and construction industries.[9] Geotextiles are being widely used over the world for road construction for providing durable and sustainable roads with better riding surface.[10] Among the different geotextiles, Jute Geotextile (JGT) provides indigenous, available technologies, which have got enough potentialities for improvement of all types of paved roads.[11] Although several varieties of JGT both woven and non-woven have since been developed for a number of geotechnical end uses, it is a fact that there is an urgent need to design and develop precise fabric as overlay on existing pavements and other emerging civil engineering applications for its technically and economically competitive nature in our country as well as in global market. Hence there is an ample scope for doing scientific research studies on application of suitable overlay jute based bitumen impregnated paving fabric, which should be aimed at developing newer products with acceptable geotechnical properties. It is in this context that Central Road Research Institute (CRRI), New Delhi, India in collaboration with Department of Jute and Fibre Technology, University of Calcutta, West Bengal, India has carried out a thorough scientific research on the development of Bituminized Jute Paving Fabric which will serve as a partial substitute of bitumen mastic reinforcing the pavement. The CRRI scientists have undertaken several laboratory simulation tests for assessing the fabric's functional performance for the said applications for each of the developed overlay jute composite paving fabrics under suitable conditions that can resist abrasion of moving loads expected on different categories of roads (e.g. State High Ways, Major District Roads and Rural Roads, etc.) and standardization of the same. Throughout the tenure of this research work all of the properties of the different grades and types of bitumen and their compatibility with jute fibre/fabric have been studied extensively after which they have submitted their comprehensive research and development report along with recommendation on the developed fabric to be used for field trials. But this has been felt that if the properties, grades and types of bitumen are not studied thoroughly then it will be difficult for a textile engineer to have a better understanding of the bitumen impregnated different types of jute geotextiles. Therefore, in the outlook of making a textile engineer and engineers of other relevant fraternity well conversant with bitumen and most of its properties for production of bituminized jute geotextile for its wider application in the field of civil engineering, particularly in road construction, as a diversified product out of jute. This paper has tried to give an overall review of bitumen and its properties.

4.2   Bitumen

Indian Standard Institutions define Bitumen[12] as a 'black or dark brown non-crystalline soil or viscous material having adhesive properties derived from Petroleum Crude either by natural or by refinery processes'. The common binders used in bituminous road constructions are Road Tars and Bitumen. Bitumen has gradually replaced road tar for road construction purposes mainly because of its greater availability as compared to road tars.

Figure 4.1   Trinidad Lake asphalt, Venezuela.

The varieties of naturally occurring bitumen, as shown in Fig. 4.1 are asphalt and asphaltites.[13] Asphalt is also produced as a petroleum by-product. Both substances are black and largely soluble in carbon disulfide. Asphalts are of variable consistency, ranging from a highly viscous fluid to a solid, whereas asphaltites are all solid. Asphalts fuse readily, but asphaltites fuse only with difficulty. Asphalts may, moreover, occur with or without appreciable percentages of mineral matter, but asphaltites usually have little or no mineral matter.

The characteristics of naturally available bitumen as reported by Hoiberg[14] are: Bitumen is solid and semisolid hydrocarbons that can be converted into liquid form by heating. Bitumen can be refined to produce commercial products such as gasoline, fuel oil and asphalt which are shown in Fig. 4.2 and Fig. 4.3, respectively. Bitumen is a general name for various solid and semisolid hydrocarbons. In 1912, the term was used by the American Society for Testing and Materials (ASTM) to include all those hydrocarbons that are soluble in carbon disulfide, whether gases, easily mobile liquids, viscous liquids or solids. Bitumen is a generic term applied to natural flammable substances of variable colour, hardness and volatility, composed principally of a mixture of hydrocarbons substantially free from oxygenated bodies.[15] Bitumen is sometimes associated with mineral matter, the non-mineral constituents being fusible and largely soluble in carbon disulfide, yielding water-insoluble sulfonation products. Petroleum, asphalts, natural mineral waxes and asphaltites are all considered as bitumen.[16]

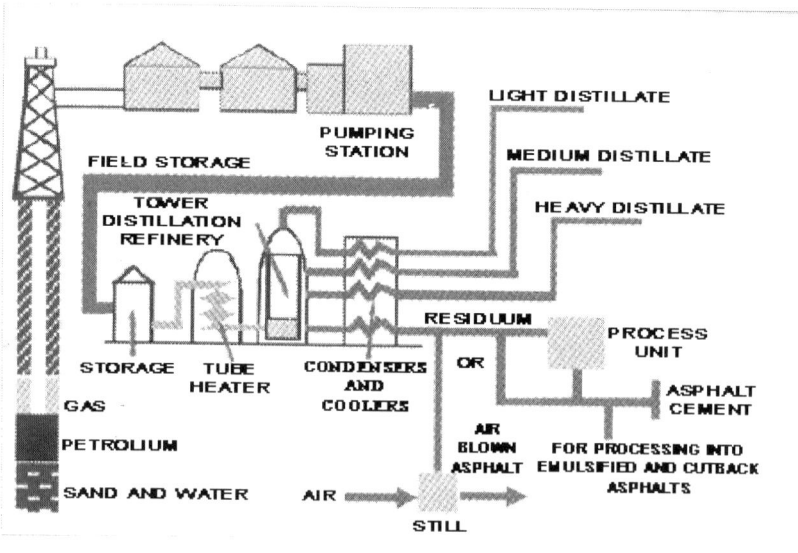

Figure 4.2   Refinery operation.

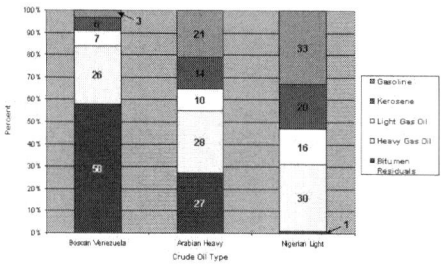

Figure 4.3 Make-up of crude oil.

Bitumen is often obtained as admixture with other materials since prehistory throughout the world for use as sealant, adhesive, building mortar, road asphalt, incense and decorative application on pots, buildings or human skin. The material is also useful in waterproofing canoes and for coating the walls of other water transport materials.

4.3 General uses of bitumen[17]

Bituminous uses can be broadly classified under two main heads:

1. Civil engineering work

(a) Construction of roads, runways, platforms

(b) Mastic floorings for factories, go downs

(c) Water proofing to prevent water seepage

2. Industries

(a) Electric cables

(b) Sealing compound in battery

(c) Bituminous felts

4.4 Types of bitumen

Bitumen or bituminous binder available in India is mainly of the following types:

4.4.1 Penetration grade[18]

- Bitumen 80/100: The characteristics of this grade confirm that of S 90 grade of IS-73-1992.[19] This is the softest of all grades available in India. This is suitable for low volume roads and is still widely used in the country.

- Bitumen 60/70: This grade is harder than 80/100 and can withstand higher traffic loads. The characteristics of this grade confirm to that of S 65 grade of IS-73-1992. It is presently used mainly in construction of National Highways & State Highways.[20]

- Bitumen 30/40: This is the hardest of all the grades and can withstand very heavy traffic loads. The characteristics of this grade confirm to that of S 35 grade of IS-73-1992. Bitumen 30/40 is used in specialized applications like airport runways and also in very heavy traffic volume roads in coastal cities in the country.

4.4.2 Industrial grade bitumen

Industrial grade bitumen is also known as blown bitumen. This is obtained by blowing air into hot bitumen at high temperatures (normally beyond 180°C). Blowing hot air into bitumen at high temperatures results in structural changes in bitumen. Esters are formed in this process and these esters link up two different molecules and higher molecular weight material increases drastically. In the process the asphaltene content is increased which in turn results in higher softening points and very low penetration number. Industrial grade bitumen is used in industrial applications and in water proofing, tar felting, etc.[21]

4.4.3 Cutback

Cutback is a free flowing liquid at normal temperatures and is obtained by fluxing bitumen with suitable solvents. The viscosity of bitumen is reduced substantially by adding kerosene or any other solvent. Cutback has been used in tack coat applications.

4.4.4 Bitumen emulsion

Bitumen emulsion is a free flowing liquid at ambient temperatures, which is a stable dispersion of fine globules of bitumen in continuous water phase. Dispersion is obtained by processing bitumen and water under controlled conditions through a colloidal milk together with selected additives. The use of proper quality emulsifiers is essential to ensure that the emulsion has stability over time and also that it breaks and sets when applied on aggregates/ road surface. It is chocolate brown free flowing liquid at room temperature. Bitumen emulsions can be of two types cationic and anionic. Anionic bitumen emulsions are generally not used in road construction as generally siliceous aggregate is used in road construction. Anionic bitumen emulsions[23] do not give good performance with siliceous matter whereas cationic bitumen emulsions give good performance with these aggregates. Therefore, cationic bitumen emulsions are far more popular than anionic bitumen emulsions.

Emulsified Bitumen Grades[22, 23]:

(a) Rapid Setting-1 (RS-1)

(b) Rapid Setting-2 (RS-2)

(c) Medium Setting (MS)

(d) Slow Setting-1 (SS-1)

4.4.5 Modified bitumen

Modified bitumen is bitumen with additives. These additives help in further enhancing the properties of bituminous pavements. Pavements constructed with modified bitumen last longer which automatically translates into reduced overlays. Pavements constructed with modified bitumen can be economical if the overall lifecycle cost of the pavement is taken into consideration.

IS 15462:2004 classifies the polymer and rubber modified bitumen into the following four types:

(a) PMB(P)   Plastomeric thermoplastic based

(b) PMB(E)   Elastomeric thermoplastic based

(c) NRMB   Natural rubber and SBR latex based and

(d) CRMB   Crumb rubber/treated crumb rubber based

Types of Bitumen Modifiers:

The modifier should be compatible with bitumen to achieve the required properties.

Table 4.1   Classification of rubber and polymer based bitumen modifiers (IRC: SP: 53-2010)

Types of modifiers	Examples
Plastomeric thermoplastics	Polyethylene (PE), ethylene vinyl acetate(EVA), ethylene butyl acrylate (EBA), ethylene-methyl-acrylate copolymers (EMA), etc.
Elastomeric thermoplastics	Styrene isoprene styrene (SIS), styrene-butadiene-styrene (SBS) block copolymer, styrene-butadiene rubber (SBR) latex
Synthetic rubber latex	Styrene-butadiene rubber (SBR) latex
Natural rubber	Latex or rubber powder
Crumb rubber or treated crumb rubber	Crumb rubber, treated crumb rubber

4.4.6 Viscosity grade bitumen

The new method of grading the product has now rested on the viscosity of the bitumen (at 60°C and 135°C).

Table 4.2 The new viscosity grades of bitumen with nomenclature (IRC: 111-2009)

Grades	Minimum of absolute viscosity at 60°C, poises, min	Kinematic viscosity at 135°C, CST, min	Approx. penetration grade	General applications
VG 10	800	250	80–100	Use in spraying applications and for paving in very cold climate in lieu of old 80/100 grade
VG 20	1600	300	60–80	Use for paving in cold climatic, high altitude regions of North India
VG 30	2400	350	50–70	Use for paving in most of India in lieu of old 60/70 grade
VG 40	3200	400	40–60	Use in highly stressed areas— toll booths, truck parking lots

4.5   Present grading system of bitumen

Paving grade bitumen is the bitumen obtained from refineries and conforms to IS 73. Recently, the third revision of Indian Standards for Paving Bitumen Specifications IS 73:2006[20] has been released by Bureau of Indian Standards. Three grades of bitumen confirming to IS 73: 1992[19] are manufactured in India. In this third revision grading of bitumen is changed from penetration grade to viscosity grade. To improve the quality of bitumen, BIS revised IS-73-1992 Specifications based on viscosity grade (viscosity @ 60°C) in July 2006. As per the Specifications, there are four grades VG-10, VG-20, VG-30 and VG-40.

With the current revision several key issues are addressed, like:

(1) Performance at high temperatures by adopting a viscosity-graded bitumen specification (based on viscosity at 60°C), in place of the current penetration-graded specification (based on penetration at 25°C)

(2) Issues relating to compaction, which the tender asphalt mixtures create as push and shove under the roller wheels, have also addressed by having a requirement of minimum viscosity at 135°C, it will be helpful in minimizing the tender mix problems in the field.

(3) Adoption of viscosity-graded paving bitumen specifications will also reduce the number of total tests to 7.

Viscosity grades of bitumen are categorized according to viscosity (degree of fluidity) grading. The higher the grade, the stiffer is the bitumen. In viscosity grade, viscosity tests are conducted at 60°C and 135°C, which represents the temperature of road surface during summer and mixing temperature, respectively. The penetration at 25°C, which is annual average pavement temperature, has been also retained in specifications.

Table 4.3 AASHTO M 226 and ASTM D 3381 viscosity grades

Standard	Grading based on original asphalt (AC)						Grading based on aged residue (AR)				
AASHTO M 226	AC-2.5	AC-5	AC-10	AC-20	AC-30	AC-40	AR-10	AR-20	AR-40	AR-80	AR-160
ASTM D 3381	AC-2.5	AC-5	AC-10	AC-20	AC-30	AC-40	AR-1000	AR-2000	AR-4000	AR-8000	AR-16000

4.6 Properties of bitumen

4.6.1 Bitumen: a viscoelastic material

The properties of bitumen can be defined in terms analogous to the modulus of elasticity of solid materials. In case of solids, modulus of elasticity E is defined by Hooke's law. Bitumen is a viscoelastic material. At high temperatures it behaves like a liquid and hence liquid flow properties like viscosity are exhibited. However, at low temperatures bitumen behaves like a solid and hence solid properties like stress and strain become relevant. Similarly, for shorter loading time bitumen behaves like a solid whereas for longer loading times bitumen behaves like a liquid. The properties that bitumen exhibits in the intermediate temperature range and loading time are of great relevance as this range is very long and bitumen is handled in this temperature range most of the times. Due to the viscoelastic nature of bitumen, there is always a phase lag in stress and strain in case of repetitive loadings. For purely elastic material the phase lag is 0° and for purely viscous material the phase lag is 90°. In case of bitumen since it is neither a liquid nor a solid at most temperatures hence the phase lag is always in between 0° and 90°. A graphical correlation between the phase angle and modulus of a bitumen, popularly known as "Black Curve" is shown below followed by its physical interpretation.

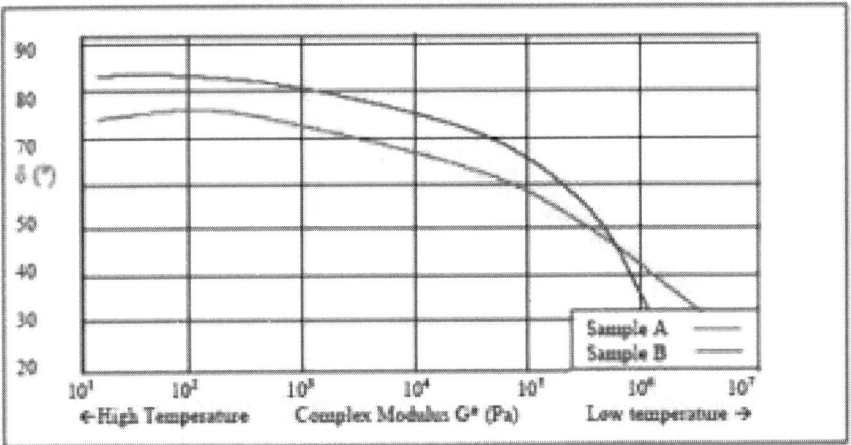

Figure 4.4   Plot of black curve phase angle (δ) vs. complex modulus (*G**).

4.6.1.1 Interpretation of results

In the results shown, sample A is a polymer modified bitumen and generally has a lower phase angle than an unmodified product, sample B. The lower the phase angle, the more elastic natured the asphalt binder will be and so it is less likely to form wheel ruts on a road in hot weather (modulus shown c. 105). At very low temperatures (modulus c. 107), the polymer modified sample has a higher phase angle than the unmodified material, indicating that it is less likely to show cracking. At very high temperatures (modulus 101) the binder softens enough to become viscous so that it can be easily mixed and laid in position. The above theory is extremely useful in studying fatigue characteristics, properties of creep and also tensile strength of bitumen.

4.6.2 Adhesion properties of bitumen

Bitumen has excellent adhesive qualities provided the conditions are favourable. However in presence of water the adhesion does create some problems. Most of the aggregates used in road construction possess a weak negative charge on the surface. The bitumen aggregate bond is because of a weak dispersion force. Water is highly polar and hence it gets strongly attached to the aggregate displacing the bituminous coating.

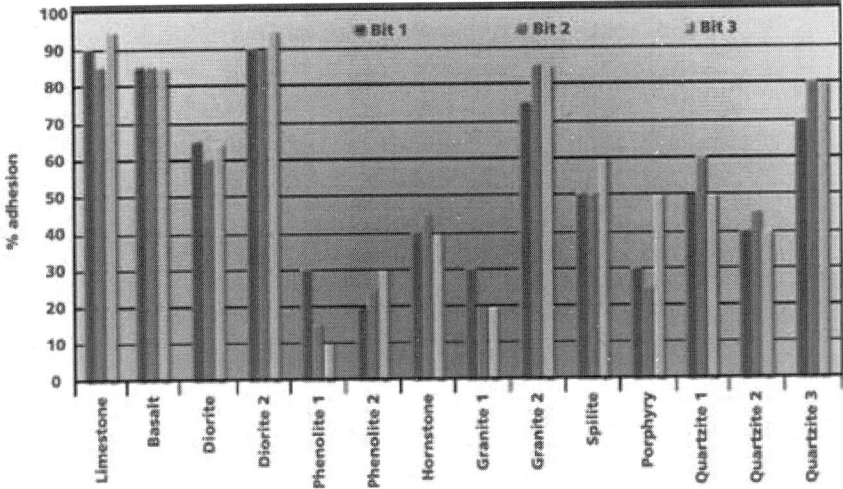

Figure 4.5   Adhesion of bitumen to minerals.

The factors influencing aggregate bitumen adhesion are plenty and some of the factors influencing this property are as below:

- External: Rainfall, humidity, water pH, presence of salts, temperature, temperature cycle traffic, design, workmanship and drainage.

- Aggregate: Mineralogy, surface texture, porosity, dirt, durability, surface area, absorption, moisture content, shape, weathering.

- Bitumen: Rheology, constitution.

- Mix: Void content, permeability, bitumen content, bitumen film thickness, filler type, aggregate grading and mix type.

4.7 Tests for bitumen

There are many bitumen properties which can be tested. All these tests replicate the actual field conditions in different ways. Different types of standard tests conducted on it are briefly described below:

4.7.1 Viscosity test[23]

The actual tests conducted are as follows:

Viscosity at 135°C is a fair indicator of the ability of bitumen to coat the aggregates properly. In order to get best coating the viscosity has to be optimum. Too viscous bitumen would result in inadequate and non-uniform coating of the aggregates. Very low viscosity would again result in inadequate coating as the bitumen will tend to bleed. Therefore, viscosity at 135°C is a true reflection of the quality of bond that is likely to be formed with the aggregate. Various testing equipments like capillary viscometer, cup viscometer, tar viscometer, etc. can be used for testing the viscosity.

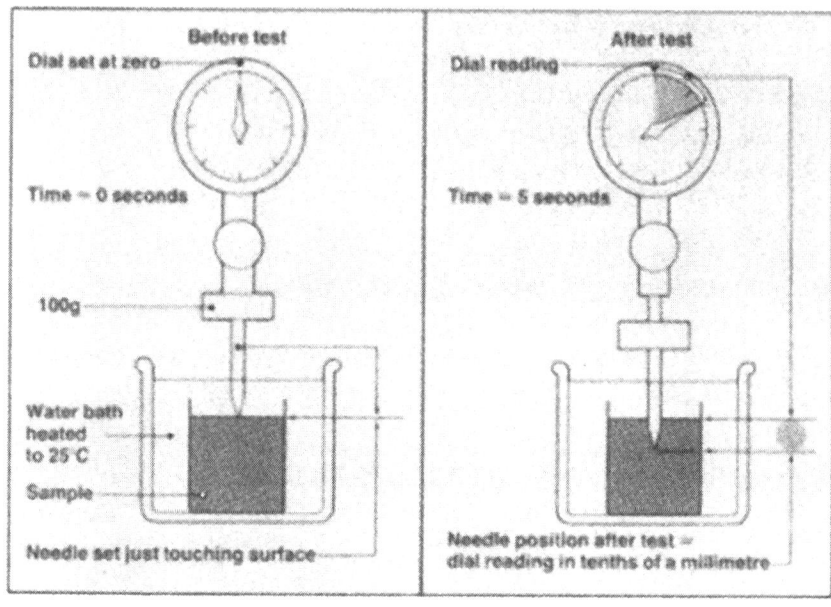

Figure 4.6   Principle of typical viscometer.

The Rotational Viscometer (RV) is used to test high temperature viscosities [the test is conducted at 135°C (275°F)]. The basic RV test measures the torque required to maintain a constant rotational speed (20 RPM) of a cylindrical spindle while submerged in an asphalt binder at a constant temperature. This torque is then converted to a viscosity and displayed automatically by the RV.

Figure 4.7   Rotational viscometer.

The RV high-temperature viscosity measurements are meant to simulate binder workability at mixing and laydown temperatures. The goal is to ensure the asphalt binder is sufficiently fluid for pumping and mixing hence a maximum RV viscosity is specified. The RV is more suitable than the capillary viscometer (used for kinematic viscosity) for testing modified asphalt binders because some modified asphalt binders (such as those containing crumb rubber particles) can clog the capillary viscometer and cause faulty readings. The standard rotational (or Brookfield) viscometer test is AASHTO TP 48 and ASTM D 4402: Viscosity determination of asphalt binder using rotational viscometer.

4.7.2 Softening point[24]

Bitumen does not have a distinct melting point. It gradually softens when heated. As there is no distinct melting point therefore the softening point test has been developed to arbitrarily indicate the transition temperature. The softening point is also an empirical test and denotes the temperature at which bitumen would behave more like a liquid and less like a solid under standard conditions of heating and loading. In this test a standard Ring and Ball Apparatus is used. The sample is taken in a standard mould and standard weights (in the form of steel balls) are placed on it. The system is then heated in a water bath at a standard rate. The temperature at which the bitumen coated steel ball touches the bottom of the beaker is called the Softening Point temperature.

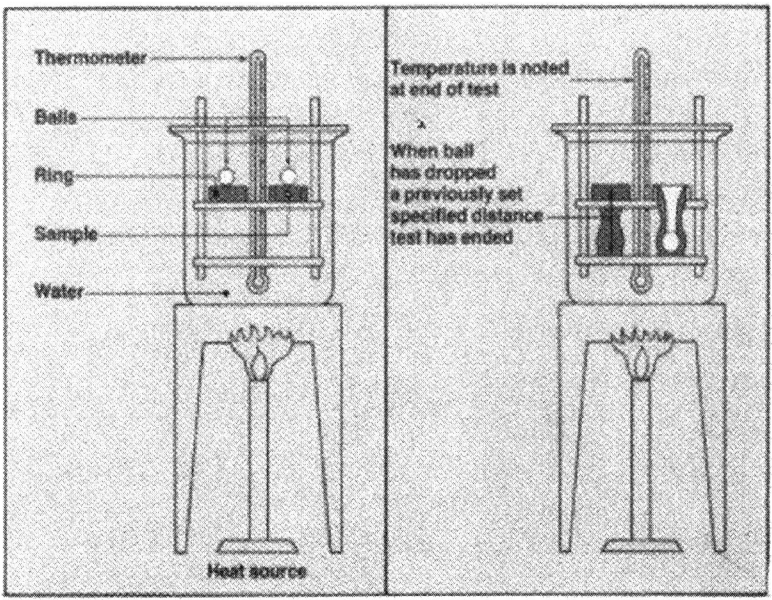

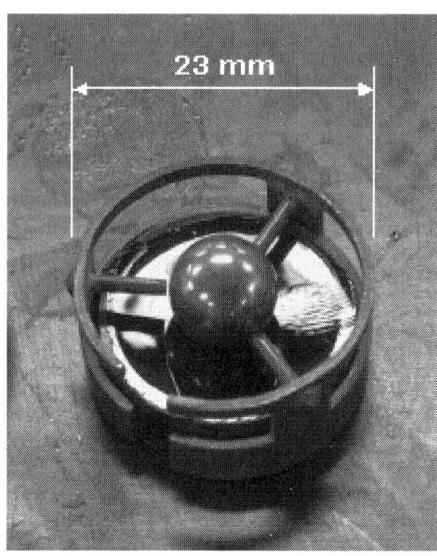

Figure 4.8   Principle of softening point test.

The standard softening point test is AASHTO T 53 and ASTM D 36: Softening point of bitumen (Ring-and-Ball Apparatus) softening point test is a very important test as it is a fair indicator of melting properties of bitumen. Bitumen with lower softening point tends to melt on the road in summer and start flowing under the impact of temperature and traffic. Subsequently when the bitumen cools down at night the road surface loses its original shape and becomes wavy. This mode of failure of roads due to bitumen is referred to as failure by rutting. Therefore, it can be concluded that bitumen with higher softening point melt at higher temperatures has better rutting resistance. The softening point serves to gauge the uniformity of supply and on account of its rapidity and accuracy, is used extensively for purposes of factory control.

4.7.3 Ductility test[25]

The ductility test is an empirical test which measures the cohesive strength of bitumen. In this test a standard size bitumen sample is maintained at a constant temperature. The sample is pulled at a constant rate at constant temperature. The length at which the sample breaks is called the ductility of the sample. One unique feature of ductility test is that the test temperature at times varies from country to country and also from grade to grade.

Ductility test is an indicator of the cohesive strength of bitumen which in turn is a very loose indicator of the fatigue strength of the material. Material with higher ductility is more likely to withstand repeated cycles of loading and unloading in a better way.

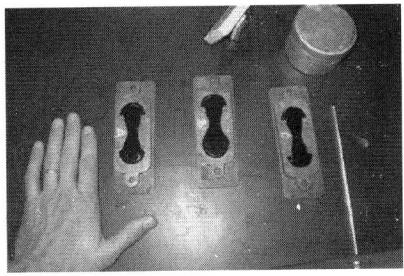 

Figure 4.9 Principle of ductility test.

The standard ductility test is AASHTO T 51 and ASTM D 113: Ductility of bituminous materials.

4.7.4 Penetration test[26]

The consistency of bitumen is determined by penetration test. Various types and grades of bituminous materials are available depending on the origin and refining process. This test measures the depth (in the units of one-tenth of a millimetre or one-hundredth of a centimetre) to which a standard needle will penetrate vertically under specified conditions of standard load, duration and temperature.

The penetration test is widely used world over for classifying the bitumen into different grades.

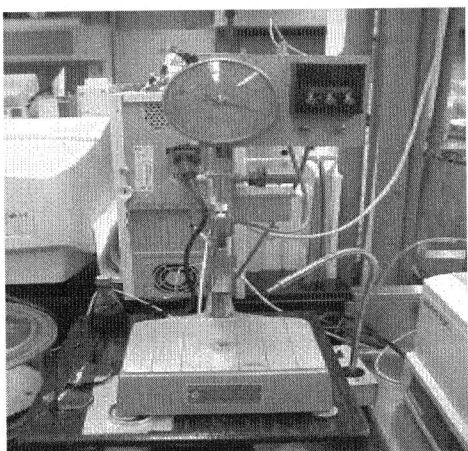

Figure 4.10 Principle of penetration test.

The basic principle is to measure the penetration (in units of one-tenth millimetre) of a standard needle in a bitumen sample maintained at 25°C for 5 s. The total weight of the needle assembly was 100 g. The softer the bitumen, the greater will be the penetration.

4.7.5 Penetration index or penetration ratio

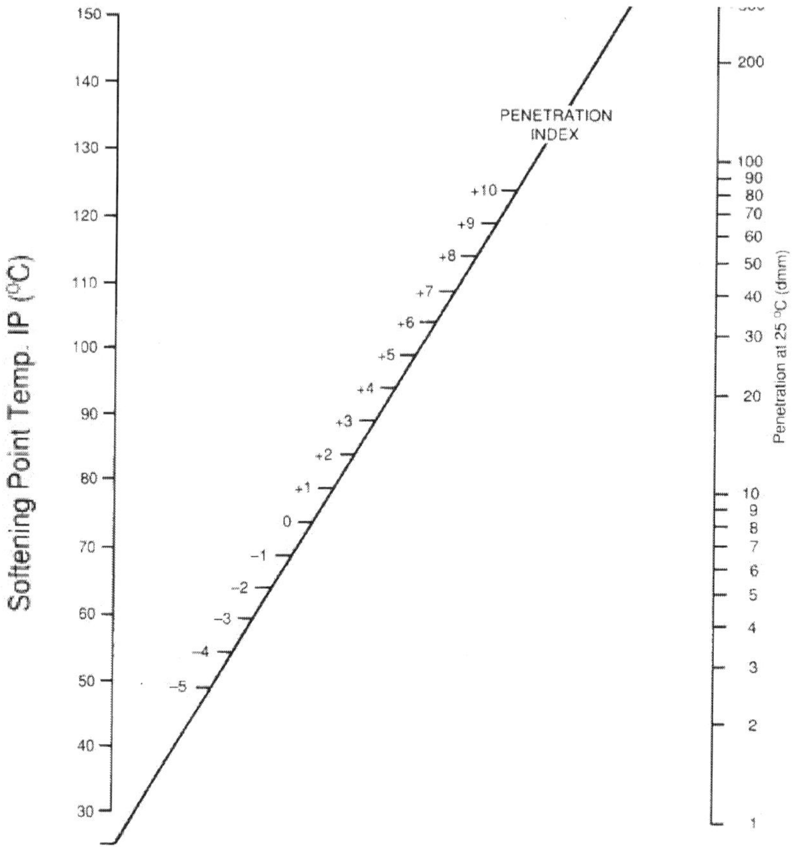

Figure 4.11   Penetration index.

The penetration of the same sample of bitumen can be measured at different temperatures and a temperature vs. penetration graph can be plotted on a log graph sheet. The graph is a straight line and the slope of this straight line is called the Penetration Index.

Penetration index (PI) can also be calculated with the help of the following formula:

$$PI = 20(1-25A)/(1 + 50A)$$

The value of A (and PI) can be derived from penetration measurements at two temperatures, T_1 and T_2, using the equation

$$A = \log (\text{penetration at } T_1) - \log (\text{penetration at } T_2)/T_1 - T_2$$

Penetration is related to viscosity and empirical relationships have been developed for Newtonian materials. If penetration is measured over a range of temperatures, the temperature susceptibility of the bitumen can be established.

The consistency of bitumen may be related to temperature changes by the expression

$$\log P = AT + K$$

where,

P = penetration at temperature T

A = temperature susceptibility

K = constant

Penetration index is a fair indicator of the ability of bitumen to resist repeated variations in the temperature of the pavement. Penetration ratio is a simplified version of the penetration index. It is very similar to penetration index but in this case the sample is tested with 100 g weight on the needle at 25°C and 200 g weight on the needle at 4°C. While deriving the values of penetration index and penetration ratio the assumption is that the properties of bitumen vary in a linear manner over the entire range of temperature (in service as well during application).[27]

4.7.6 Matter soluble in organic solvents

This test measures the presence of inorganic impurities in bitumen. Solvents like trichloroethylene, carbon disulphide, carbon tetrachloride, toluene, etc. are used for this purpose. In this test bitumen is dissolved in the solvent (trichloroethylene, carbon disulphide, carbon tetrachloride or toluene) and the material insoluble in the solvent is filtered out. It is then repeatedly washed with the solvent to remove all soluble matter. The insoluble matter that is finally left behind is weighed and the percentage calculated. The choice of solvent has been a matter of debate and discussion in the scientific community. Some of these solvents are considered to be toxic and hazardous. The laboratories and test method specification making bodies prefer not to use these toxic solvents and have switched over to less toxic or non-toxic solvents.

4.7.7 Rotating thin film oven test (RTFOT)

Once the bitumen is found to be meeting the viscosity criterion the next step of aging the sample in the laboratory is undertaken. The conventional TFOT test is replaced by rotating thin film oven test. In the RTFOT small bottles, like medicine bottles, are coated with bitumen on the inner side and the bottles are fixed in the oven on a shelf in horizontal position. A jet of air is periodically blown into each bottle to speed up the oxidation process. Therefore this test is faster test and can cause aging equivalent to two years (after laying) within 135 min.

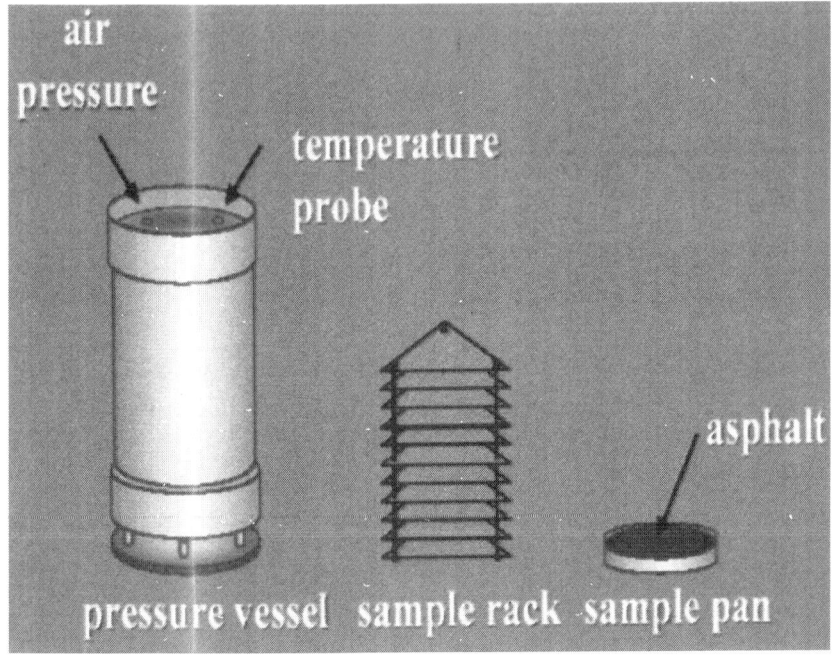

Figure 4.12   Rotating thin film oven test (RTFOT)—assembly.

The rolling thin-film oven (RTFO) test simulates short-term aging by heating a moving film of asphalt binder in an oven for 85 min at 163°C (325°F). The effects of heat and air are determined from changes incurred in physical properties measured before and after the oven treatment by other test procedures.[28] The moving film is created by placing the asphalt binder sample in a small jar then placing the jar in a circular metal carriage that rotates within the oven. The RTFO test is generally considered superior to the TFO because: It achieves the same degree of hardening (aging) in less time (85 min vs. 5 h), it uses a rolling action that allows continuous exposure of fresh asphalt binder to heat and air flow which allows asphalt binder modifiers, if used, to remain dispersed in the sample and prevents the formation of a surface skin on the sample, which may inhibit aging.

The standard RTFO test is AASHTO T 240 and ASTM D 2872: Effects of heat and air on a moving film of asphalt (rolling thin-film oven test).

RTFO Arrangement

RTFO Samples (left – after aging in the RTFO, centre – before aging in the RTFO, right – empty sample jar)

Figure 4.13   Rolling thin-film oven test apparatus.

4.7.8 Pressure aging vessel (PAV)

The PAV was adopted to simulate the effects of long-term asphalt binder aging that occurs as a result of 5 to 10 years HMA. The PAV was adopted to simulate the effects of long-term asphalt binder aging that occurs as a result of 5 to 10 years HMA pavement service (Bahia and Anderson, 1994). The general concept of the pressure aging vessel had been used for many years in rubber product aging. The PAV is an oven-pressure vessel combination that takes RTFO aged samples) and exposes them to high air pressure [2070 kPa

(300 psi)] and temperature [90°C (195°F), 100°C (212°F) or 110°C (230°F) depending upon expected climatic conditions] for 20 h. Aging the asphalt binder samples under pressure is advantageous because—there is a limited loss of volatiles and the oxidation process can be accelerated without resorting to extremely high temperatures.

PAV Sample

Figure 4.14   Pressure aging vessel.

Standard PAV test is AASHTO *PP1: Practice for accelerated aging of asphalt binder using a pressurized aging vessel.*

4.7.9 Flash point

The flash point tests the flammability of bitumen. Asphalt cement like most other materials, volatilizes (gives off vapour) when heated.[29] At extremely high temperatures (well above those experienced in the manufacture and construction of HMA) asphalt cement can release enough vapour to increase the volatile concentration immediately above the asphalt cement to a point where it will ignite (flash) when exposed to a spark or open flame. This is called the flash point. For safety reasons, the flash point of asphalt cement is tested and controlled. The fire point, which occurs after the flash point, is the temperature at which the material (not just the vapours) will sustain combustion. A typical flash point test involves heating a small sample of asphalt binder in a test cup. The temperature of the sample is increased and at specified intervals a test flame is passed across the cup. The flash point is the lowest liquid temperature at which application of the test flame causes the vapours of the sample to ignite. The test can be continued up to the fire point—the point at which the test flame causes the sample to ignite and remain burning for at least 5 s.

Clevel.and open cup tester

Pensky martens closed cup tester

Figure 4.15 Different flash point testers.[18]

Standard flash point tests are (1) AASHTO T 48 and ASTM D 92: Flash and fire points by Cleveland open cup (more common for asphalt cement used in HMA) and (2) AASHTO T 73 and ASTM D 93: Flash-point by Pensky–Martens closed cup tester.

4.7.10 Specific gravity test[19]

The density of a binder is a fundamental property frequently used as an aid in classifying the binder for use in paving jobs. The specific gravity is greatly influenced by the chemical composition of the binder. Increased amount of aromatic type compounds cause an increase in the specific gravity, the test procedure have been standardized by the ISI. The specific gravity of bituminous material is defined by ISI as 'the ratio of the mass of the given volume of bituminous material to the mass of an equal volume of water, the temperature being specified as $(27 \pm 0.1)°C$'.

Formula for calculating the specific gravity of bitumen:

Weight of specific gravity bottle = $W_1 g$

Weight of specific gravity bottle filled with water = $W_2 g$

Weight of specific gravity bottle half filled with bitumen = $W_3 g$

Weight of specific gravity bottle half filled with bitumen and rest half filled with distilled water = $W_4 g$

Specific gravity of bitumen = (Weight of bitumen)/(Weight of equal volume of water) = $(W_3 - W_1)/(W_2 - W_1) - (W_4 - W_3)$

4.7.11 Dynamic shear rheometer (DSR)[30]

The DSR is used for testing medium to high temperature viscosities [the test is conducted between 46°C (115°F) and 82°C (180°F)]. The actual temperatures anticipated in the area where the asphalt binder will be placed determine the test temperatures used.

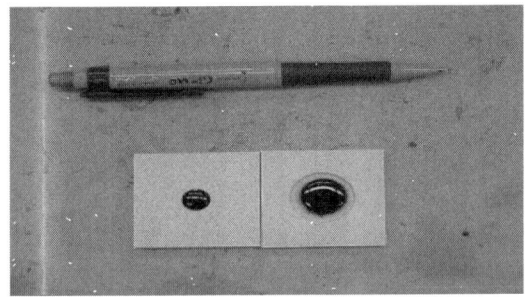

Figure 4.16   Dynamic shear rheometer (DSR).

$$\tau_{max} = \frac{2T}{\pi r^3} \circ \quad \gamma_{max} = \frac{\theta r}{h} \circ \quad G^* = \frac{T_{max}}{\gamma_{max}} \circ \quad \delta = \text{time lag}$$

The basic DSR test uses a thin asphalt binder sample sandwiched between two plates. The lower plate is fixed while the upper plate oscillates back and forth across the sample at 1.59 Hz to create a shearing action. These oscillations at 1.59 Hz (10 radians/s) are meant to simulate the shearing action corresponding to a traffic speed of about 90 km/h (55 mph) (Roberts et al., 1996). The following equations are then used to determine a complex shearing modulus, G^* and a phase angle.

where, τ_{max} = maximum applied shear stress

T = maximum applied torque

r = radius of binder specimen (either 12.5 or 4 mm)

γ_{max} = maximum resulting shear strain

θ = deflection (rotation) angle

h = specimen height (either 1 or 2 mm)

G^* = complex shear modulus

δ = Phase angle. This is the time lag (expressed in radians) between the maximum applied shear stress and the maximum resulting shear strain. For typical neat asphalt (no modifiers) the phase angle is about 88–89°, while some modified binders have phase angles as low as 60°.

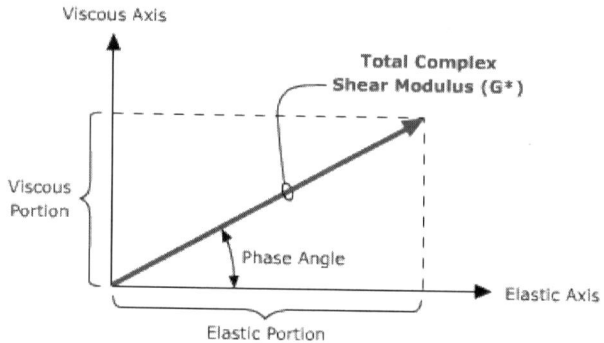

Figure 4.17   Complex shear modulus components.

Asphalt binders in the medium to high temperature range behave partly like an elastic solid (deformation due to loading is recoverable—it is able to return to its original shape after a load is removed) and a viscous liquid (deformation due to loading is non-recoverable—it cannot return to its original shape after a load is removed). By measuring G^* and δ, the DSR is able to determine the total complex shear modulus as well as its elastic and viscous components.

4.7.12 Bending beam rheometer (BBR)

The BBR is used to test asphalt binders at low temperatures where the chief failure mechanism is thermal cracking. The BBR basically subjects a simple asphalt beam to a small (100-g) load over 240 s. Then, using basic beam theory, the BBR calculates beam stiffness $[S(t)]$ and the rate of change of that stiffness (m-value) as the load was applied.

$$S(t) = \frac{PL^3}{4bh^3\delta(t)}$$

where,	$S(t)$	=	creep stiffness at time, $t = 60$ s
	P	=	Applied constant load (980 ± 20 mN), obtained using a 100 g load. Note that 100 g multiplied by the force of gravity (9.8 m/s²) = 0.98 N, or 980 mN
	L	=	distance between beam supports, 102 mm
	b	=	beam width, 12.5 mm
	h	=	beam thickness, 6.25 mm
	δ	=	deflection at time, $t = 60$ s

Figure 4.18   Bending beam rheometer.

The *m*-value is simply the rate of change of the stiffness at time, $t = 60$ s and is used to describe how the asphalt binder relaxes under load.

Sl. No.	Property	Test method	Test result	Specified values
1.	Penetration, 0.1 mm, 25°C	IS:1203-1978 ASTM D5	63	60–70
2.	Softening point, °C	IS:1205-1978 ASTM D36	49	40–55

3.	Ductility, cm	IS:1208-1978	+75	75 (min)
		ASTM D113		
4.	Specific gravity, 27°C	IS:1202-1978	1.01	0.99 (min)
		ASTM D70		
5.	Flash point, °C	ASTM D92	250°C	250°C

4.8 Failure mechanism of asphalt

Robertson et al. (1991) describe asphalt behaviour in terms of its failure mechanisms. They describe each particular failure mechanism as a function of asphalt's basic molecular or intermolecular chemistry. This section is a summary of Robertson et al. (1991).

4.8.1 Aging[31]

Some aging is reversible, some is not. Irreversible aging is generally associated with oxidation at the molecular level. This oxidation increases an asphalt's viscosity with age up until a point when the asphalt is able to quench (or halt) oxidation through immobilization of the most chemically reactive elements.[31] Reversible aging is generally associated with the effects of molecular organization. Over time, the molecules within asphalt will slowly reorient themselves into a better packed, more bound system. This results in a stiffer, more rigid material. This thixotropic aging can be reversed by heating and agitation. Bitumen, like any organic matter, is affected by factors like presence of oxygen, ultraviolet rays and changes in temperature. These factors are responsible for hardening of bitumen. Hardening results in decrease in penetration increase in softening point and increase in penetration index (PI). For increased life of bituminous pavement it is essential that excessive hardness does not take place. Hardening of bitumen takes under the influence of external factors in the following ways:

4.8.2 Oxidative hardening

When bitumen is exposed to atmosphere for a prolonged period the oxygen starts reacting with the bitumen constituents and higher molecular weight molecules are formed. Larger molecules result in lesser flexibility and hence increased hardness. The degree of hardness is dependent on factors like ambient temperature, exposure time and thickness of bitumen film. It is observed that for 10°C increase in temperature above 100°C the oxidation rate doubles.

4.8.3 Hardening due to loss of volatiles

Over a period of time the volatile components in bitumen evaporate. The rate of evaporation is dependent on temperature only. The volatiles in bitumen are relatively very low and hence hardening due to loss of volatiles is relatively small.

4.8.4 Physical hardening

At ambient temperatures bitumen molecules slowly reorient themselves. This results in physical hardening. This process is an extremely slow process and hence actual hardening due to the above factor is very low.

4.8.5 Educative hardening

Educative hardening takes place due to the movement of oily components out of bitumen over a period of time. The rate of hardening due to this process is dependent on the type of bitumen and also on the porosity of the aggregate.

4.8.6 Hardening of bitumen during storage

Hardening of bitumen during storage can be easily minimized by taking a few simple precautions. Bitumen is stored in above ground tanks at high temperatures and high temperature and presence of oxygen are the two primary factors responsible for hardening of bitumen. Hence it is very important that bitumen be handled at the lowest possible temperature, consistent with efficient use. Also the storage tanks should have low surface to volume ratio so as to minimize the exposed surface area. Lower exposed surface area would mean lower oxidation rate.

4.8.7 Hardening of bitumen on road

Some hardening of bitumen can take place on the road also due to oxidation. The level of oxidation is purely dependent on the access to oxygen. If the pavement is well graded and well compacted the hardening is nominal as the void content will be low.

4.8.8 Rutting and permanent deformation

If the molecular network is relatively simple and not interconnected, asphalt will tend to deform inelastically under load (e.g. not all the deformation is recoverable). Additionally, asphalts with higher percentages of non-polar dispersing molecules are better able to flow and plastically deform because the various polar molecule network pieces can more easily move relative to one another due to the greater percentage of fluid non-polar molecules.[32]

4.8.9 Fatigue cracking

If the molecular network becomes too organized and rigid, asphalt will fracture rather than deform elastically under stress. Therefore, asphalts with higher percentages of polar, network-forming molecules may be more susceptible to fatigue cracking.

4.8.10 Thermal cracking

At lower temperatures even the normally fluid non-polar molecules begin to organize into a structured form. Combined with the already-structured polar molecules, this makes asphalt more rigid and likely to fracture rather than deform elastically under stress.

4.8.11 Stripping

Asphalt adheres to aggregate because the polar molecules within the asphalt are attracted to the polar molecules on the aggregate surface. Certain polar attractions are known to be disrupted by water (itself a polar molecule). Additionally, the polar molecules within asphalt will vary in their ability to adhere to any one particular type of aggregate.

4.8.12 Moisture damage

Since it is a polar molecule, water is readily accepted by the polar asphalt molecules. Water can cause stripping and/or can decrease asphalt viscosity. It typically acts like a solvent in asphalt and results in reduced strength and increased rutting.32 When taken to the extreme, this same property can be used to produce asphalt emulsions. Interestingly, from a chemical point-of-view water should have a greater effect on older asphalt. Oxidation causes aged (or older) asphalts to contain more polar molecules. The more polar molecules asphalt contains, the more readily it will accept water. However, the oxidation aging effects probably counteract any moisture-related aging effects.

4.9   Modified bitumen[33]

4.9.1 Advantages of modified bitumen

(a)   Lower susceptibility to temperature variations.

(b) Higher resistance to deformation/wear and tear.

(c) Better adhesion between aggregates and binder.

(d) Increase in fatigue life.

(e) Resistance in reflective cracking.

(f) Better age resistance properties.

4.9.2 Types of modified bitumen

A variety of additives are used for modification of bitumen. The degree of modification depends on type of modifier, its dose and nature of bitumen. The most commonly used modifiers are:

4.9.2.1 Synthetic polymers—plastomeric thermoplastics

(a) Low Density Polyethylene (LDPE)

(b) Ethylene Vinyl Acetate (EVA)

(c) Ethylene Butyl Acetate (EBA)

(d) Ethylene Tar Polymer (ETP)

4.9.2.2 Synthetic polymers—elastomeric thermoplastics

(a) Styrene Isoprene Styrene (SIS)

(b) Styrene Butadiene Styrene Block Copolymer

4.9.2.3 Natural rubber

(a) Latex Powder

(b) Rubber Powder

4.9.2.4 Crumb rubber

(a) Crumb Rubber without additives

(b) Crumb Rubber with additives

4.10 Different types and grades of bitumen

(a) Industrial available bitumen types are as follows:

(i) Bitumen 30/40 (ii) Bitumen 60/70 (iii) Bitumen 80/100

(b) Different types of modified and emulsified bitumen which are commercially available

(1) Polymer Modified Bitumen (Plastomeric and Elastomeric)—PMB 40, PMB 70, PMB 120

(2) Crumb Rubber Modified Bitumen—CRMB 50, CRMB 55, CRMB 60

(3) Cationic Bitumen Emulsions—RS_1, RS_2, SS_1, SS_2, MS

4.11   Bitumen requirements for various applications

Bitumen is used in different forms like Tack Coat, Bituminous Macadam and Bitumen Mastic in road construction as per the specifications given below:

Sl. No.	Category	Quantity (kg/10 m²)
1.	Tack Coat	2.0–2.5
	Normal bituminous surface	2.5–3.0
	Dry bituminous surface	2.5–3.0
	Granular surface (with primer)	3.5–4.0
	(d) Non-bituminous surface	
2.	Bituminous Macadam	
	Compacted thickness (50 mm)	50
	Compacted thickness (75 mm)	68
3.	Bitumen Mastic	14–17

Figure 4.19 Freshly placed emulsion tack coat. The brown colour indicates that it has not yet broken.

Figure 4.20 The same tack coat after 23 min. The brown colour now appears in splotches indicating its break.

Figure 4.21 Tack coat using an asphalt emulsion. The black colour indicates it has broken.

4.12 Health, safety and environmental aspects

Bitumen presents a low order of potential hazard provided that good handling practices are observed. Hence, it is absolutely necessary that adequate safety precautions are taken while handling bitumen.[34] In case of accidental contact with hot bitumen and affected part should be immediately plunged in water. Ice pack can also be given. However, no attempt should be made to remove firmly adhered bitumen from the skin. It can be allowed to fall off gradually or can be removed by medicinal paraffin. In all cases the effected person should be taken to qualified doctor immediately. During mixing of bitumen with aggregate, fumes are emitted. These fumes contain particulate matter, hydrocarbon vapours and very small amount of H_2S. However, the concentration is rarely above permissible limits. Bitumen also contains polycyclic aromatic hydrocarbons (PCA). PCAs with molecular weight of 200 to 450, especially benzo (∞) pyrene, are carcinogenic. Other than heat burn, hazards associated with skin contact of most bitumens are negligible. However, it is prudent to avoid prolonged and intimate skin contact.

References

1. Raizada, P.S. and Tikare, P., Safe roads: A mission impossible? Indian Highways, 40(6), 41, June 2012.

2. Ling, H.I. and Liu, Z., Performance of geosynthetic-reinforced asphalt pavements, Geotechnical and Geoenvironmental Engineering, 127(2), 2001.

3. *Guidelines on use of Modified Bitumen in Road Construction*, Second Revision, IRC:SP:53-2010.

4. Dondi, G., Full scale dynamical testing on reinforced bituminous pavements. In: Proc. Geosynthetics'97, Industrial Fabrics Association International, Roseville, Minn., pp. 749–762, 1997.

5. Haas, R., Structural behavior of tensar reinforced pavements and some field applications, Polymer Grid Reinforcement, London, pp. 166-170, 1985.

6. Koerner, R.M., Designing with Geosynthetics, fourth edition, Prentice Hall, Upper Saddle River, NJ, 1998.

7. Baker, T., The most overlooked factor in paving fabric installation, Geotechnical Fabrics Report, 48–52, 1998.

8. Barazone, M., Installing paving synthetics—an overview of correct installation procedures, Geotechnical Fabrics Report 17–20, 2000.

9. Holtz, R.D., Christopher, B.R. and Berg, R.R., Geosynthetic Engineering, Bi Tech Publishers Ltd., Canada, 1997.

10. Ingold, T.S., The Geotextiles and Geomembrane Manual, Elsevier Advanced Technology, UK, 1994.

11. Meccai, K.A. and Hasan, E.A., Geotextiles in Transport Applications, Second Gulf Conference on Roads, Abu Dhabi, UAE.

12. Don, L.L., Physical Geology, sixth edition, Prentice-Hall, Englewood Cliffs, NJ, 1982.

13. Abraham, H., Asphalts and Allied Substances, sixth edition, Vol. 5, 1960–1963.

14. Hoiberg, A.J., Bituminous Materials: Asphalts, Tars, and Pitches, Vol. 1, 1964.

15. Mathew, V. and Krishna Rao, K.V., Pavement materials: bitumen. Introduction to Transport Engineering, NPTEL, Chapter 23, 8 May 2007.

16. Durand, G. and Piorier, J.E., Particle size effects in bitumen emulsions. In: AEMA International Symposium on Asphalt Emulsions, Washington DC, 1996.

17. Booth, E.H., Gaughan, R.G. and Holleran, G., Some uses of cationic emulsions in NSW and South Australia Part 2. In: Australian Road Research Board Int. Conf. Perth, 1994.

18. The design and use of granular emulsion mixes. Division of Roads and Transport Technology CSIR, GEMS–SABITA Manual, SABITA Ltd., Roggebaai, 14 October 1993.

19. Paving Bitumen Specifications, Third Revision, IS: 73-2006.

20. Paving Bitumen Specifications, IS: 73-1992.

21. American Association of State Highway and Transportation Officials (AASHTO).

22. Nösler, I., A contribution to the objective and quantitative measurement of the adhesion between aggregates and bitumen. Thesis Work, University of Wuppertal, Institute of Road Construction, 1999.

23. Kandhal, P.S., An overview of the viscosity grading system adopted in India for paving bitumen, Indian Roads Congress, Indian Highways, April 2007.

24. Standard test method for softening point of bitumen (Ring-and-Ball Apparatus) ASTMD36-95. In: Annual Books of ASTM Standards, Philadelphia, PA, Vol. V04.04, 1998.

25. Goodrich, J.L., Asphalt and polymer modified asphalt properties related to the performance of asphalt concrete mixes. In: Proc. AAPT, vol. 57, 1988.

26. Wami, E.N., Puyate, Y.T. and Chigeru, S.N., Characterization of penetration-grade bitumen blended with lighter petroleum products, Global Journal of Engineering Research, 7(1), 1–6, 2008.

27. Mashaan, N.S., Asim Ali, H., Rehan Karim, M. and Abdelaziz, M., Effect of crumb rubber concentration on the physical and rheological properties of rubberised bitumen binders, International Journal of the Physical Sciences, 6(4), 684-690, 18 February 2011.

28.   Pfeiffer J.Ph., The Properties of Asphaltic Bitumen, Elsevier Publishing Company, Inc., Amsterdam, Netherlands, p. 161, 1950.

29. Hveem, F.N., Quality tests for asphalts—a progress report. In: Proceedings of the Association Asphalt Paving Technologists, Vol. 15, p. 111, 1943.

30. Standard test method for determining the rheological properties of asphalt binder using a dynamic shear rheometer. American Association of State Highway and Transportation Officials AASHTO provisional standard TP5-93: AASHTO, Washington DC, 1995.

31. Herold, M. and Roberts, D., Spectral characteristics of asphalt road aging and deterioration: Implications for remote-sensing applications, Applied Optics, 44(20), 4327–4334, 2005.

32. Fwa, T.F., Tan, S.A. and Zhu, L.Y., Rutting prediction of asphalt pavement layer, *Journal of Transportation Engineering*, 130(5), 675–683, 2004.

33. Bitumens and bitumen derivatives. In: Conservation of Clean Air and Water in Europe CONCAWE (PD 92/104). Brussels, Belgium, (Sensors 2008) 81293, December 1992.

34. Lee, D.Y. and Demirel, T., Beneficial effects of selected additives on asphalt cement mixes. Final Report. ISU-ERI-Ames-88070 Report Number IDOT Project HR-278. Ames, Iowa Engineering Research Institute, Iowa State University, 1988.

35. *Chu, V.T.H., A Self Learning Manual—Mastering Different Fields of Civil Engineering Works (VC-Q-A-Method).*

A comprehensive idea about paving fabric

5.1 Introduction

Geotextiles are textiles applied in soil to help its engineering performance.[1] Man-made geotextiles are made of artificial fibres like polypropylene, polyethylene and some other petrochemical derivatives.[2] Natural geotextiles, on the other hand, are made out of natural fibres like jute, coir, sisal, etc.[3] To make effective use as an overlay fabric on existing pavements, paving fabric has to be waterproof and abrasion resistant.[4] Extensive research works are going on to make a durable and cost-effective smooth road transport system.[5] The research works are mainly based on construction of roads by using suitable geotextiles. Application of geotextiles in flexible paved road construction is an established one and is increasing at rapid pace throughout the world.[6] Geotextiles extend the service life of roads, increase their load carrying capacity and reduce rutting.[7] The effectiveness of geotextiles in stabilization and separation roles with flexible pavements has been extensively researched. It has been found that for weak sub-grade (CBR = 2%), the geotextile extends the service life of a flexible pavement section by a factor of 2.5–3.0 compared to a non-stabilized section.[8] Further a geotextile effectively increased the pavement section's total AASTHO structural number by approximately 19%.[9] The effect of geotextile in pavement sections with moderate strengs (CBR = 4.2–4.5%) is that the geotextile increased the service life of the pavement section by a factor of 2.0–3.3 and the AASTHO structural number increased by 13–22%.[10] These significant improvements are obtained primarily through the separation function of the geotextile placed at the interface of the base course aggregate and sub-grade soil. Asphalt concrete pavement overlays can benefit from the use of paving fabric interlayer.[11] The documented field experience indicates to number of positive benefits which includes waterproofing of the lower layers, thereby maintaining higher material strengths, retarding reflection cracking in the overlay by acting as a Strain Absorbing Membrane Interlayer (SAMI) as shown in Fig. 5.1, increase in structural stability by providing for more stable sub-grade moisture contents.[12]

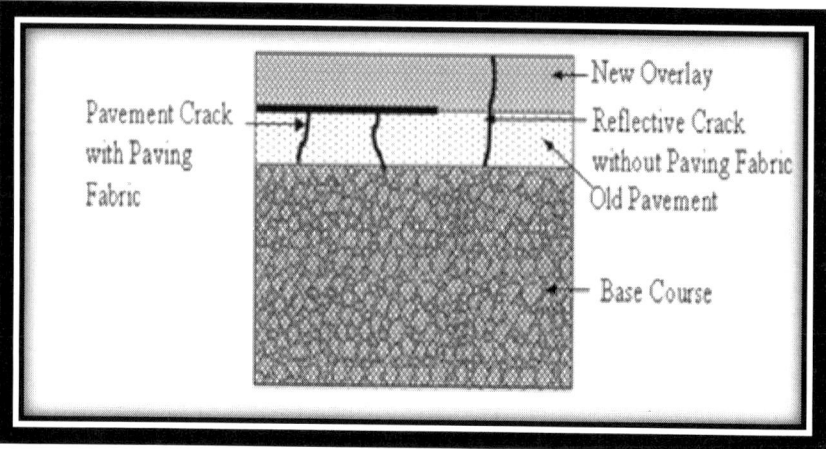

Figure 5.1   Application of Strain Absorbing Membrane Interlayer (SAMI).

Paving fabrics can also be used in new pavements to provide the same benefits. If fabric is added and the overlay thickness is not reduced from that determined by normal methods, then an increase in performance can be obtained.[13] The inclusion of a non-woven paving fabric interlayer system significantly improves the performance of asphalt concrete overlays. This performance improvement is a result of both the water proofing capabilities and the stress absorption capabilities of the paving fabric system. Synthetic fabrics and SAMI have been effective in controlling low to medium severity alligator cracking. They may be also useful for controlling reflection of temperature cracks when used in combination with crack filling. They generally do little, however, to retard reflection of cracks subjected to significant horizontal or vertical movement.[14] Introduction of paving fabrics to asphalt overlay reinforcement paved road surfaces must be maintained when they develop significant cracks and potholes.[15] The rehabilitation of cracked roads by simple overlaying, or placing an additional layer of asphalt over the old paved surface is rarely durable solution. The cracks in the old pavement eventually propagate through the new surface.[16] This is called reflective cracking. In spite of limited resistance to reflective cracking traditional overlays are still the most common approach to maintaining distressed pavements.[17] Generally, the thicker the overlay the longer it will last, however, the thicker overlays are more expensive.[18] Many research works have been done to find out the performance of paving fabrics as an interlayer to retard reflective cracks.[19] They have found that a stress relieving geosynthetic interlayer retards the development of reflective cracks by absorbing the stresses that arises from the damaged surface.[20] Reinforcement occurs when a geosynthetic is able to contribute significant tensile strength to the pavement system.[21] The reinforcement attempts to prevent the cracked old pavement from moving under traffic loads and thermal stresses by holding the cracks together. Newly manufactured geocomposites can provide both stress relief and reinforcement. The benefits of geosynthetic interlayers include saving to 1–2 in. of overlay thickness, delaying the appearance of reflective cracks, lengthening the useful life of the overlay. Most of the works regarding introduction of paving fabrics in road constructions are dealt with geosynthetic, which consists of woven, non-woven and combination of both woven and non-woven fabrics. Use of natural fibre like jute is rare to find though it is available in abundant. Allover the world, asphalt concrete overlays on existing pavements are often used as a cost-saving treatment for cracked pavements. A major problem encountered with asphalt resurfacing is the phenomenon of reflective cracking and is one of the most important degradation modes in pavements. According to Amini,[22] it may be caused by shear and tensile stresses in the asphalt layer, induced by traffic loads, change in temperature, expansive sub-grade soils, moisture changes, existing cracks and crack movements in the underlying pavement. Some of the treatment techniques used to reduce reflective cracking include increased thickness of asphalt overlay, asphalt binder modification and the use of stress absorbing interlayers such as paving fabrics. The non-woven geotextile interlayer system, known as paving fabrics, may be used to reinforce asphalt overlays and delay crack reflection. It has been reported that non-woven jute geotextile (JGT) is an extremely good receptor of hot bitumen, besides having thermal compatibility with bitumen in the range of 190°C.[23] In Brazil, following an international trend, the field performance of overlays using fabric interlayers has generally been successful, although there have been cases where the paving fabric systems provided little or no improvements. Even in successful cases, it has been difficult to quantify the efficiency of those systems, especially due to the necessity of monitoring pavement performance and the acting traffic for many years. In 1999, the Pavement Laboratory of the Federal University of Rio Grande do Sul (LAPAV/UFRGS) and the Roads Department (DAER) of that Southern Brazil state started a long-term research, supported by a geotextile manufacturing enterprise, with the purpose of comparatively evaluating the performances of asphalt overlays on cracked pavements with and without using paving fabrics. Two kinds of geotextiles were used as intermediate layers in a test section loaded by a linear simulator and in a real time loading road section. Non-woven paving fabrics have been used in asphalt overlays for over four decades. Even though these non-woven geotextiles have been widely and successfully used in a variety of design and construction situations, the applications of paving fabric have been generally characterized by limited documentation of their performance. In addition, the present pavement structural design methods do not provide a rational basis for assessing or evaluating the benefits of inclusion of paving fabric in a pavement structural section. These two factors, i.e. limited performance data and lack of design approach, have resulted in a lack of consistent application of paving fabrics, and significant regional differences in the usage level of non-woven paving fabric. A Paving Fabrics Task Group was formed within the Geotextile Division of the Industrial Fabrics Association International (IFAI) to summarize the uses, benefits and applications of non-woven geotextiles for pavement membrane interliner systems, commonly referred to as paving fabrics. In 1997, Maxim Technologies Inc. presented the final report of a study undertaken to define applications, performance, benefits, extent of usage of non-woven paving fabrics and to recommend design criteria for its application in pavement management systems.[24]

The literature review examined more than 200 reports on the use and performance on non-woven fabric interlayer systems. This data collection also reviewed a database of over 100 pavement sections on which performance of the system was monitored. The data search found 4 principal end-use applications of the non-woven paving fabric system which are: paving fabric with a chip seal over unpaved roads and sub-grades, paving fabric and chip seal over existing AC (Asphalt Concrete) pavements, paving fabric and AC overlay over existing PCC (Portland Cement Concrete) pavements and paving fabric and AC overlay over AC pavements. Performance of non-woven fabric is often measured by comparing the condition of overlays using fabric with overlays not using fabric. This performance is sometimes reported in two ways. First, a comparison of the performance of paving fabric/overlay section with section containing no paving fabric (control section) is done. Second, the performance of the paving fabric section with paving fabric is compared with section with no paving fabric section but with a thicker asphalt overlay.

Carmichael and Marienfeld evaluated the application of paving fabrics in the rehabilitation of existing AC pavements in thirty sites, twenty-one within the United States and nine others are in South Africa, Germany, Spain, Belgium and Austria. Almost half of the sites were evaluated for more than 5 years and only two of the sites were observed for 10 or more years.[25] Almost all of the case studies found that the use of a paving fabric in the rehabilitation of AC pavements resulted in a significant retardation of reflection cracking. In the four case studies that indicated no real difference in reflection cracking between control and fabric sections the sections were thin overlays of less than 4.0 cm. One section was exposed to one of the most severe winters on record and one section was placed on a pavement with sub-grade depressions and extensive transverse, longitudinal and alligator cracking. The report of Maxim Technologies Incorporation states that AC overlays can benefit from the use of paving fabric interlayers.[24] The documented field experience indicates to a number of positive benefits including waterproofing of the lower layers, thereby maintaining higher materials strength. By maintaining a lower moisture content in the road base materials the effective strength or support provided by that road base is improved. Well drained (dry) bases provide up to

2.5 times the pavement support of poorly drained (wet) pavements. Retarding reflection cracking in the overlay by allowing for a stress absorbing interlayer reduces strains. An installed asphalt saturated non-woven fabric layer allows for slight differential movements between the top of the old pavement and the bottom of the AC overlay. This retards the development of reflective cracking since some of the movement associated with old cracking will be absorbed by the paving fabric layer and not transferred up into the overlay. The creation of a layered pavement system with a paving fabric interlayer also minimizes harmful tensile stresses.

The effect of thickness of overlay has been studied by several investigators. In the conclusions of the report of Maxim Technologies Inc. it is stated that the paving fabric system gives additional overlay performance equivalent to increased overlay thickness of 2.50–4.50 cm with an average performance equivalency of approximately 3.30 cm. The paving fabric interlayer is not meant to be a structural layer to make up pavement structural thickness deficiencies. Also, this amount of AC overlay equivalency does assume that the existing pavement is stable, the paving fabric is properly installed, and a minimum overlay thickness of 4.0 cm is placed.

Banerjee et al.[26] made an attempt to develop asphalt overlay fabric from jute to retard crack propagation in the asphalt overlay. The design concept of developing asphalt overlay fabric of moderate capability made purely from jute suitable for reinforcement and moisture barrier functions for low traffic roads were taken for consideration. A leno based woven construction was selected for the purpose in order to obtain breaking strength of 30–40 kN/m and extension at peak load less than 5% in both warp and weft directions. The new developed Jute Asphalt overlay (JAO) fabric within pavement was investigated. To judge the performance of the JAO fabric, they made asphalt concrete beams (ACB) with and without asphalt overlay (A/O) fabrics and the beams were subjected to cyclic mechanical loading in accelerated mode simulating the vehicular traffic on MTS (Material Testing System) to monitor the crack propagation. Both dry and hygrally treated samples of ACBs without and with A/O fabrics were subjected to cyclic mechanical loading. Under all experimental conditions they found that ACBs embedded with JAO do not exhibit any crack propagation beyond the level at which the JAO is placed within the ACB. They explained it, that JAO having grid like structure with suitable opening size help in creating proper interlocking among aggregates of the overlay and voids of the old pavement surface areas the fabric within pavement and thereby the two layers of the pavement act as single body which further resists crack growth.

The dimension of the asphalt concrete beam of their laboratory experiment was 225 mm (L) × 75 mm (W) × 75 mm (H). A transverse notch of 5 mm depth and 3 mm base width at the beam centre was created to simulated preexisting crack in the pavement. Asphalt overlay fabric reinforcement was placed at 20 mm above the beam specimen base for reinforcement specimen. The tensile properties of JAO fabrics which they have developed to find in-situ performance are given in Table 5.1.

Table 5.1 Tensile properties of JAO fabrics

Type of test	Type of sample	Breaking load kN (CV%)		Breaking elongation % (CV%)		Young's modulus in MPa (CV%)	
		Warp way	**Weft way**	**Warp way**	**Weft way**	**Warp way**	**Weft way**
Grab (kN)	JAO (double layer)	0.98 (10.2)	0.84 (10.9)	9.4 (7.1)	9 (5.6)	104.9 (8.9)	96.1 (9.3)
Wide width (kN/m)	JAO (double layer)	24.2 (14.1)	24.8 (3.3)	5.1 (13.4)	4.9 (6.8)	79.6 (7.7)	107.0 (14.0)
	JAO (double layer)	38.7 (1.7)	36.3 (4.8)	4.9 (6.2)	5.0 (3.0)	222.0 (6.1)	185.7 (7.9)

5.2 Pavements

Pavements are civil engineering structures used for the purpose of operating wheeled vehicles safely and economically. Paved roadways that include the carriageways and the shoulders have been constructed for more than a century. Their basic design methods and construction techniques have undergone some changes, but the development of geosynthetics in the past three decades had provided the strategies for enhancing the overall performance of the paved roadways. Various levels of government, in most of the countries devote unprecedented time and resources to roadway construction, maintenance and repair. Commonly a paved road becomes a contender for maintenance when its surface shows significant cracks and potholes. Cracks in the pavements surface cause numerous problems, including riding discomfort for users, reduction of safety, infiltration of water and subsequent reduction of the bearing capacity of the sub-grade, pumping of soil particles through crack, progressive degradation of the road structure in the vicinity of the cracks due to stress concentrations. The construction of asphalt overlays[27] is the most common way to renovate both flexible and rigid pavements. Most overlays are done predominantly to provide a water proofing and pavement crack retarding treatment. A minimum thickness of the asphalt concrete overlay may be required to provide an additional support to a structurally different pavement. An asphalt overlay is at least 25 mm thick and it is placed on top of the distressed pavement. Overlays are economically practical, convenient and effective. The cracks under the overlay rapidly propagate through the new surface which is a major drawback of asphalt overlays.

5.3 Paving fabrics

Many pavements that are considered to be structurally sound after the construction of a new overlay, prematurely exhibit a cracking pattern similar to that which existed in the underlying pavement. These types of cracks destroy surface continuity, decrease structural strength and allow water to enter the pavement layers.[28] Gradually these cracks propagate from underlay to overlay which exposes the inability of the overlay to withstand shear and tensile stresses created by movements of the underlying pavements due to either traffic loading or by thermal ingress and thermal effects. Thorough research and extensive studies on such structural stability problems of the roads, in isolation, have led to the adoption of paving fabrics accordingly. Paving fabrics, which are used in asphalt concrete overlay, are geosynthetics, which are classified into woven and non-woven. This type of fabric has relatively high elongation and low tensile strength. This is commonly used for stress relief.[29] When impregnated with tack coat the paving fabric allows considerable movement around a crack but nullifies or at least lessens the effect, the movements have on the overlay. These types of interlayers also waterproof the road structure. Review says that paving fabrics were first used in 1930s when cotton sheets were installed as reinforcement to asphalt layers in North Carolina, USA. Since the 1970s the concept of geotextile reinforcement of surfacing seals has been used successfully internationally with hundreds of millions of square metres installed worldwide. The paving fabric shall meet the physical requirements[30] (Table 5.2).

Table 5.2 Physical requirements—paving fabrics

Property	Units	Standard requirements	Test method
Tensile strength	kg	36.3	ASTM D4595
Grab strength	N	450.0	ASTM D 4632
Ultimate elongation	%	≥ 50	ASTM D4632
Mass per unit area	g/m^2	140	ASTM D5261

Asphalt retention	kg/10 m²	10	Texas DOT 3099
	l/m²	Notes: (3) and (4)	ASTM D 6140
Melting point	°C	150	ASTM D276
Surface texture	–	Heat bonded on one side only	Visual inspection

Notes:

1. All numerical values represent minimum average roll values (average of test results from any sampled roll in a lot shall meet or exceed the minimum values) in weaker principal direction. Lot shall be sampled according to ASTM D 4354, 'Practice for sampling of Geosynthetics for Testing'.

2. Conformance of paving fabrics to specification property requirements shall be determined as per ASTM D 4579, 'Practice for Determining the Specification Conformance of Geosynthetics'.

3. Asphalt is required to saturate paving fabric only. Asphalt retention must be provided in manufacturer certification. Value does not indicate the asphalt application rate required for construction.

4. Product asphalt retention property must meet the MARV value provided by the manufacturer certification.

5.4   Functions of paving geosynthetics

A geosynthetic layer, especially a geotextile layer, is used beneath asphalt overlays, ranging in thickness from 25 to 100 mm of AC and PCC paved roads. The geotextile layer is generally combined with asphalt sealant, or tack coat to form a membrane interlayer system known as a paving fabric interlayer. Fig. 5.2 shows the layer arrangement in paved roads with a paving fabric interlayer. When properly installed, a geotextile layer beneath the asphalt overlay mainly function31 as—fluid barrier (if impregnated with bitumen, that is, asphalt cement), protecting the underlying layers from degradation due to infiltration of road-surface moisture; and cushion, that is, stress-relieving layer for the overlays, retarding and controlling some common types of cracking, including reflective cracking. A paving fabric, in general, is not used to replace any structural deficiencies in the existing pavement. However, the above functions combine to extend the service life of overlays and the roadways with reduced maintenance cost and increased pavement serviceability.

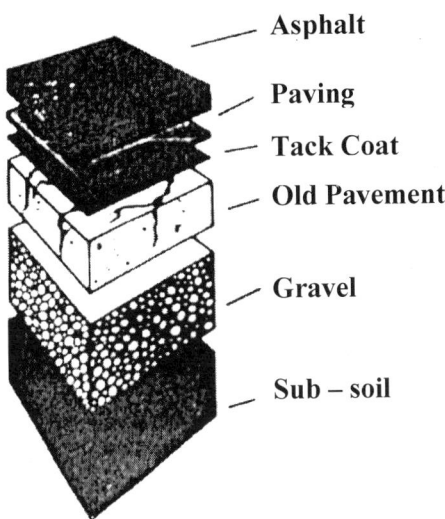

Figure 5.2　Typical cross-section of a paved road with paving fabric interlayer.[32]

The pavements typically allow 30–60% of precipitation to infiltrate and weaken the road structure. The fluid barrier function of the bitumen-impregnated geotextile may be of considerable benefit if the sub-grade strength is highly moisture sensitive. In fact, excess moisture in the sub-grade is the primary cause of premature road failures. Heavy vehicles can cause extensive damage to roads, especially when the soil sub-grade is wet and weakened. The pore water pressure can also force the soil fines into the voids in the sub-base/base, weakening them if a geotextile is not used as a separator/filter. Therefore, efforts should be made to keep the soil sub-grade at fairly constant and low moisture content by stopping moisture infiltration into the pavement and providing proper pavement drainage. A stress-relieving interlayer[27] retards the development of reflective cracks in the overlay by absorbing the stresses induced by underlying cracking in the old pavement. The stress is absorbed by allowing slight movements within the paving fabric interlayer inside the pavement without distressing the asphalt concrete overlay significantly. In fact, the addition of a stress relieving interlayer reduces the shear stiffness between the old pavement and the new overlay, creating a buffer zone that gives the overlay a degree of independence from movements in the old pavement. Pavements with paving fabric interlayers also experience much less internal crack developing stress than those without. This is why fatigue life of a pavement with a paving fabric interlayer is many times that of a pavement without, as shown in Fig. 5.3. A stress-relieving interlayer also waterproofs the pavement, so when cracking does occur in the overlay, water cannot worsen the situation.

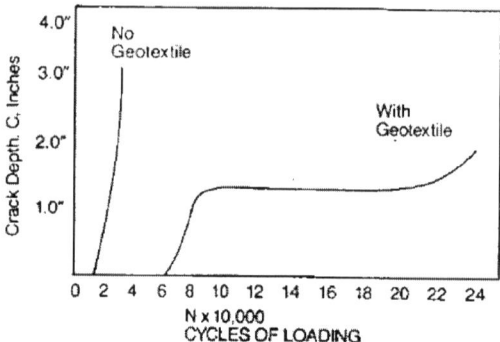

Figure 5.3   Fatigue response of asphalt overlay.[33]

Geotextiles generally have performed best when used for load-related fatigue distress, e.g. closely spaced alligator cracks. Fatigue cracks, mainly caused by too many flexures of the pavement system, should be less than 3 mm wide for best results. Geotextiles used as a paving fabric interlayer to retard thermally induced fatigue cracking,[27] caused by actual expansion and contraction of underlying layers, mostly within the overlay, have, in general, been found to be ineffective. For getting the best results on existing cracked pavement, the geotextile layer is laid over the entire pavement surface or over the crack, spanning it by 15–60 cm on each side, after placement of an asphalt levelling course followed by an application of tack coat, and then asphalt overlay is placed above as shown in Fig. 5.1. This construction technique is adopted keeping in view that much of the deterioration that occurs in overlays is the result of unrepaired distress in the existing pavement prior to the overlay. The selection of a geosynthetic for use in asphalt overlays is complicated by the variable condition of the existing roadway systems. The deterioration may range from simple alligator cracking of the pavement surface to significant potholes caused by failure of the underlying sub-grade. It is important to note that an overlay system as well as a paving fabric interlayer will fail if the existing deficiencies in the existing pavements are not corrected prior to the placement of overlay and/or paving fabric. The selected paving grade geosynthetic must have the ability to absorb and retain the asphalt tack coat, sprayed on the surface of the old pavement, to effectively form a permanent fluid barrier and cushion layer. The most common paving grade geosynthetics are lightweight needle-punched non-woven geotextiles, with a mass per unit area of 120–200 g/m^2. Woven geotextiles are ineffective paving fabrics because they have no interior plane to hold asphalt tack coat and so do not form an impermeable membrane. They also do not perform well as stress-relieving layer to help reduce cracking. Tests should be performed to determine the bitumen retention of paving fabrics for their effective application. In the most commonly used test procedure, after taking weights individually, test specimens are submerged in the bitumen at a specified temperature, generally 135°C for 30 min. Specimens are then hung to drain in the oven at 135°C for 30 min from one end and also 30 min from the other end to obtain a uniform saturation of the fabric. Upon completion of specimen submersion in bitumen, and draining, the individual specimens are weighed and asphalt/bitumen retention, R_B, is calculated by using the equation[33] $-R_B = (W_{sat} - W_f)/_B \times A_f$, where W_{sat} is the weight of saturated test specimen in kg; W_f is the weight of paving fabric in kg; A_f is the area of fabric test specimen in m^2 and $_B$ is the unit weight at 21°C in kg/l. The average bitumen retention of specimens is calculated and reported in l/m^2. Paving fabrics pre-coated with modified bitumen are also available commercially in the form of strips. These products perform the same functions of waterproofing and stress relief as the field impregnated paving fabrics; however, they are more expensive. Their applications are economical if only limited areas of the pavement need a paving fabric interlayer system. For waterproofing and covering the potholes, the pre-coated paving fabrics are good.

Heavy-duty composites of geosynthetic and bituminous membrane are commercially used, especially over cracks and joints of PCC pavements that are overlaid with asphalt concrete.

5.5 Design aspects of geosynthetics

The fluid barrier function of geosynthetic should be achieved in field application, keeping the fact in view that the water (coming from rain, surface drainage or irrigation near pavements), if allowed to infiltrate into the base and sub-grade, can cause pavement deterioration by softening the soil sub-grade, mobilizing the soil sub-grade into the road base stone, hydraulically breaking down the base structures, including stripping bitumen-treated bases and breaking down chemically stabilized bases. The selected paving grade geosynthetic should meet the physical requirements described in Table 5.1. Prior to laying paving fabric, the tack coat should be applied uniformly to the prepared dry pavement surface at the rate governed by the equation[34] $Q_d = 0.36 + Q_s + Q_c$, where Q_d is the design tack coat quantity (kg/m^2); Q_s is the saturation content of the geotextile being used (kg/m^2) to be provided by the manufacturer and Q_c is the correction based on tack coat demand of the existing pavement surface (kg/m^2). The quantity of tack coat is critical to the final membrane system. Too much tack coat will leave an excess between the fabric and the new overlay resulting in a potential sliding failure surface and potential bleeding problems, while too little will fail to complete the bond and create the impermeable membrane. In fact, the misapplication of the tack coat can make the difference between paving fabric installation success and failure. The asphalt tack coat forms a low permeability layer in the fabric and bonds the system to the existing pavement and overlay. The fabric allows slight movement of the system, while holding the tack-coat layer in place and maintaining its integrity. The actual quantity of tack coat will depend on the relative porosity of the old pavement and the amount of bitumen sealant required to saturate the paving fabric being used.[35] The quantity of sealant required by the existing pavement is a critical consideration. The saturation content of the fabric depends primarily on its thickness and porosity, which is its mass per unit area. It is to be noted that the more the mass per unit area of the geotextile, the more tack coat is required to saturate the fabric. For typical paving fabrics in the 120–135 g/m^2 mass per unit area range, most manufacturers recommend fabric-bitumen absorption of about 900 g/m^2, or application rates of about 1125 g/m^2. For the full waterproofing and stress-relieving benefits, the paving fabric must absorb at least 725 g/m^2. The remaining part of the applied bitumen helps in bonding the system with the existing pavement and the overlay. Additional tack coat may be required between the overlap to satisfy saturation requirements of the fabric.

A review of projects with unsatisfactory paving fabric system performance shows the importance of the tack coat to the whole system. Evidence from records of 65 projects, which took place over a 16-year period, indicates that the tack coat application was too light (less than 725 g/m^2) in an overwhelmingly high percentage of failure cases. This is shown graphically in Fig. 5.4. In the laboratory tests it has been observed that the waterproofing benefit of paving fabric is negligible until the fabric absorbs at least 725 g/m^2 of tack coat. Inadequate tack coat may result in rutting, shoving or, occasionally, complete delamination of the overlay. It has been found that the structural problems such as overlay slippage and delamination begins to occur where the tack coat quantity absorbed by the fabric is less than about 450 g/m^2. In addition to low application amounts of tack coat, there can be another set of conditions that may result in a low tack amount in paving fabric. Inadequate rolling or, low overlay temperatures may create conditions in which the tack may not be taken up by the fabric. In fact, overlays less than 40 mm thick are seldom recommended with paving fabric, in part, because of their rapid heat loss.

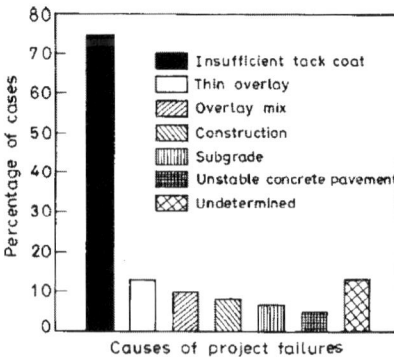

Figure 5.4   Causes in 65 project failures investigated in the United States between 1982 and 1997.

5.6 General design considerations

5.6.1 Site selection

Previous field experiences have shown that the existing pavement section should not show any sign of vertical movement. To maximize the benefit of paving fabrics, pavements must be structurally sound, with no existing surface delamination.

5.6.2 Pavement evaluation

Field evaluation should include a visual distress survey in accordance with accepted methodology. As performance is based on experimental assessments, there are no defined limits on deflection or curvature, as the fabric interlayer with its retained bitumen, encourages movement with deflection.

5.6.3 Case studies

A research project on development of bituminized jute paving fabric (BJPF) for strengthening of pavement and partial substitute of bitumen mastic sponsored by National Jute Board (NJB), Ministry of Textiles, Government of India has been carried out by Department of Jute and Fibre Technology, University of Calcutta, India in collaboration with Central Road Research Institute (CRRI), New Delhi, India. Grey jute Paving Fabric (GJPF) which is a combination of woven and non-woven jute fabric had been developed in a commercial jute mill followed by its bituminization along with the testing of the physical, mechanical and hydraulic property parameters of the developed fabric related to geotechnical applications as well as laboratory simulation testing carried out at CRRI laboratory, New Delhi. After the fabric has been developed and tested the same has been subjected to several field trials which had been executed at busy roads of different traffic intensities both at the heart as well as at the fringes of the metropolitan cities including the state highways and national highways and all of the trials are under constant monitoring conducted at regular intervals. So far the performance assessment report of the developed fabric obtained from the monitoring is quite satisfactory. Some of the major case studies of the application of the developed bituminized jute paving fabric are furnished in the subsequent topics to come.

References

1. Martin, P., Sarsby, W.R. and Anand, S.C., Hand Book of Technical Textiles, Horrocks, A.R. and Anand, S.C. (Eds.), Woodhead Publishing in association with The Textile Institute, Manchester, 2000.

2. John, N.W.M., Geotextiles, Blackie and Sons Ltd., Glasgow, 1987.

3. Ramaswamy, S.D. and Aziz, M.A., Proceedings of International Workshop on Geotextiles, Bangalore, India, pp. 259, 1989.

4. Abdullah, A.B.M., Proceedings of International Workshop on Jute Geotextiles-Technical Potential and Commercial Prospects, Kolkata, pp. 69, 2008.

5. Liekweg, M. International Fiber Journal, 19(2), 10, 2004.

6. Menon, J.P. and Basu, G.L.S., Journal of the Indian Roads Congress, 74(3), 315, 2013.

7. Sprague, C.J. and Carver, C.A., Asphalt Overlay Reinforcement-Geotechnical Fabrics Report, pp. 30, 2000.

8. Specification for Road and Bridge Works, Ministry of Road Transport & Highways, IRC, New Delhi, pp. 302, 2001.

9. The American Association of State Highway and Transportation Officials (AASTHO). www.transportation.org.

10. Aziz, M.A. and Ramaswamy, S.D. Some Studies on Jute Geotextiles and their Applications-Geosynthetic World, Wiley Eastern Limited, 1996.

11. Amini, F., Potential Applications of Paving Fabrics to Reduce Reflective Cracking. Final Report. Jackson State University, Jackson, Mississippi. www.mdot.state.ms.us/research/pdf/PavFabr.pdf, 2005.

12. Marienfeld, M.L. and Smiley, D. Paving Fabrics: the Why and the How-To-Geotechnical Fabrics Report, pp. 24, 1994.

13. Carmichael, R.F. and Marienfeld, M.L., Synthesis and Literature Review of Nonwoven Paving Fabrics Performance in Interlayers. Transportation Research Record, TRB, 1687, 112, 1999.

14. Baker, T., The Most Overlooked Factor in Paving-Fabric Installation. Geotechnical Fabrics Report, pp. 48, 1998.

15. Indian Road Congress, Tentative Specification for Slurry Seal and Micro-Surfacing, IRC: SP-81, New Delhi, India, 2008.

16. Haas, R., Structural Behaviour of Tensar Reinforced Pavements and Some Field Applications. Polymer Grid Reinforcement, London, 1985.

17. Barksdale, R.D., Brown, S.F. and Chan, F., Potential benefits of geosynthetics in flexible pavement systems. Nat. Cooperative Hwy. Res. Program Rep. Transportation Research Board, Washington D.C., 1989.

18. Dondi, G., Proceedings of Full Scale Dynamical Testing on Reinforced Bituminous Pavements. Geosynthetics 97, Industrial Fabrics Association International, Roseville, Minn., pp. 749, 1997.

19. Giroud, J.P. and Noiray, L., Journal of Geotechnical Engineering Division, ASCE, 107(9), 1233, 1981.

20. Holtz, R.D. and Sivakugan, K., Geotextiles and Geomembranes, 5(3), 191, 1987.

21. Specifications for Road and Bridge Works, IRC, New Delhi, 2001.

22. Amini, F., Potential Applications of Paving Fabrics to Reduce Reflective Cracking-Final Report. Jackson State University, Jackson, Mississippi, 2005.

23. Koerner, R.M., Designing with Geosynthetics, fourth edition, Prentice Hall Inc., New Jersey, 1997.

24. Maxim Technologies Inc., Nonwoven Paving Fabrics Study. Final Report, Submitted to the Industrial Fabrics Association International, Geotextile Division, 1997.

25. Carmichael, R.F. and Marienfeld, M.L. Synthesis and Literature Review of Nonwoven Paving Fabrics Performance in Interlayers, Transportation Research Record, TRB, 1677, 124, 1997.

26. Banerjee, P.K., Ghosh, M. and Rao, G.V., Journal of the Textile Institute, 101(5), T431, 2008.

27. Sprague, C.J. and Carver, C.A., Asphalt Overlay Reinforcement. Geotechnical Fabrics Report, pp. 30, 2000.

28. GFR, Overlay Stress Absorption and Reinforcement. Geotechnical Fabrics Report, pp. 8, 2003.

29. Peres Núñez, W., Nakahara, S. and Ceratti, J., Federal University of Rio Grande do Sul, Porto Alegre, RS, Brazil, J. Silveira, and J. Augusto de Oliveira, Rio Grande do Sul State Roads Department, Porto Alegre, Brazil Sílvio Palma, Ober S.A., Nova Odessa, SP, Brazil, The efficiency of geotextiles delaying crack reflection in asphalt mixes overlays submitted to accelerated pavement testing and real time loading, The First Pan American Geosynthetics Conference & Exhibition, 2–5 March, Cancun, Mexico, 2008.

30. Holtz, R.D., Christopher, B.R. and Berg, R.R. Geosynthetic Engineering, BiTech Publishers Ltd., Canada, 1997.

31. Barazone, M., Installing Paving Synthetics – An Overview of Correct Installation Procedures. Geotechnical Fabrics Report, pp. 17, 2000.

32. IFAI, A Design Primer, Geotextiles and Related Materials. Section 13-Asphalt Overlay, Industrial Fabrics Association International, St. Paul, USA, 1992.

33. ASTM Standard Method, To Determine Asphalt Retention of Paving Fabrics Used in Asphalt Paving for Full-Width Applications, Designation: D 6140-00, American Society of Testing Materials, West Conshohocken, USA, 2000.

34. India Roads Congress, Guidelines for Use of Geotextiles in Road Pavements and Associated Works, IRC SP-59, New Delhi, India, 2002.

35. AASHTO, Geotextile Specification for Highway Applications, AASHTO Designation: M 288-00, American Association of State Highway and Transportation Officials, Washington D.C., 2000.

6

Development of bituminized jute paving fabric (BJPF)

6.1 Introduction

Selected grey jute paving fabric (GJPF) samples, physical properties of which had been furnished in Table 3.15, Chapter 3, page 72, have been sent to Central Road Research Institute (CRRI), New Delhi, India for selection of the right grade of bitumen amongst the commercially available bitumen such as Bitumen 30/40, 60/70, 80/100, Polymer Modified Bitumen like PMB 40, PMB 70, PMB 120, Crumb Rubber Modified Bitumen CRMB 50, CRMB 55, CRMB 60 and Cationic Bitumen Emulsion MS, SS1, SS2; SPRAMUL (SS1ASTM and CQS1H) along with suitable chemical recipe for achieving the desired property parameters of the developed product.

6.2 Laboratory simulation testing in CRRI, New Delhi, India

6.2.1 Physical tests on aggregates

Aggregate forms the major part of the pavement structure as it has to primarily bear load stresses occurring on the pavement.[1] So, naturally it has to withstand the high magnitude of load stresses and wear and tear. The aggregates of different sizes (20 mm, 10 mm, 6 mm, stone dust and lime) were obtained from a hot mix plant near Delhi, India and various physical tests were carried out on them to check their suitability for use.

6.2.2 Specific gravity and water absorption test

Specific gravity of an aggregate is considered to be a measure of strength or quality of the material. Stones having low specific gravity are generally weaker than those with higher specific gravity values. The specific gravity test helps in identification of stone.[2] Water absorption gives an idea of strength of rock. Stones having more water absorption are more porous in nature and are generally considered unsuitable unless they are found to be acceptable based on strength, impact and hardness tests.[3] The test results are presented in Table 6.1. The gradation of individual aggregates is presented in Table 6.2.

Table 6.1   Test results for specific gravity and water absorption

Type of aggregates	Specific gravity	Water absorption (%)	Permissible limits as per MoRT&H, 2001
Coarse aggregates (20 mm)	2.62	0.50	
Fine aggregates (13.2 mm)	2.61	0.67	2% max.
Fine aggregates (6 mm)	2.63	0.71	
Stone dust	2.68	–	–
Lime	2.24	–	–

Table 6.2   Gradation of individual aggregates

Sieve size, mm	Percent of aggregates passing through sieve size				
	20 mm	13.2 mm	6 mm	Stone dust	Lime
26.5	100.0	100.0	100.0	100.0	100
19	65.8	100.0	100.0	100.0	100
13.2	4.7	84.1	100.0	100.0	100
9.5	0.3	24.5	96.1	98.9	100
4.75	0.0	0.5	14.5	96.2	100
2.36	0.0	0.2	0.4	81.7	100
1.18	0.0	0.1	0.3	58.8	100
0.6	0.0	0.1	0.3	48.2	100
0.3	0.0	0.1	0.3	30.9	99
0.15	0.0	0.1	0.2	18.9	89
0.075	0.0	0.1	0.1	9.7	62

6.2.3 Impact test

Toughness is the property of a material to resist impact.[4] Due to traffic loads, the road stones are subjected to the pounding action or impact and there is possibility of stones breaking into smaller pieces. The road stones should therefore be tough enough to resist fracture under impact. The impact test measures the resistance of the stones to fracture under repeated impacts. The test results are presented in Table 6.3.

Table 6.3   Test results for aggregate impact test

Type of aggregates	Aggregate impact value (%)	Permissible limits as per MoRT&H*, 2001 (for BC)
Coarse aggregates (20 mm)	19%	24% max.
Fine aggregates (10 mm)	13.35%	

MoRT&H Ministry of Road Transport and Highways.

6.2.4 Shape test

The particle shape of aggregates is determined by the percentage of flaky and elongated particles contained in it[5] [IS: 2386 (Part 1), 2002]. The presence of flaky and elongated particles is considered undesirable as they may cause inherent weakness with possibilities of breaking down under heavy loads. Angular shape is preferred due to increased stability derived from the better interlocking. The flakiness index of the aggregates is the percentage by weight of particles whose least dimension (thickness) is less than three-fifth (0.6) of their mean dimension. This test is not applicable to sizes smaller than 6.3 mm. The elongation index of the aggregates is the percentage by weight of particles whose greatest dimension is (length) greater than one and four-fifth times (1.8 times) their mean dimension. The elongation test is not applicable to sizes smaller than 6.3 mm.

6.2.5 Stripping test for aggregates

The aggregates used in bituminous pavements should have less affinity with water when compared with bituminous materials, otherwise the bituminous coating on the aggregate will be stripped off in presence of water[6] (IS: 6241, 1971). To check the stripping properties of the aggregates IS: 6241-1971 describes the procedure for stripping test. The stripping test was done on the aggregates with 60/70 bitumen. The retained coating was found to be more than 95%, which conforms to the requirements as per MoRT&H specifications, 2001 (Fourth Revision). So, the test for water sensitivity has not been carried out in the present study.

6.2.6 Physical tests on bitumen

Bitumen is a petroleum product[7] obtained by distillation of petroleum crude is used in the construction of road pavement especially in flexible pavement to withstand a relatively adverse condition of traffic and climate. Different physical tests like ductility test[8] (IS: 1208, 1978), softening point test[9] (IS: 1205, 1978), specific gravity test[10] (IS: 1202, 1978), penetration test[11] (IS: 1203, 1978) and viscosity test have been carried out. However, the impregnation of the jute samples has been done with three different binders, viz. 60/70, PMB-40 and 80/100, so the asphalt retention test has been done with all the three binders.

6.2.7 Asphalt retention testing of bitumen treated jute paving fabric

Asphalt retention is defined as the weight of asphalt cement retained by paving fabrics per unit area of specimen after submersion in the asphalt cement. The test has been done as per ASTM D 6140, 'Standard Method to Determine Asphalt Retention of Paving Fabrics used in Asphalt Paving for Full-width Applications'. The test procedure for determining asphalt retention is to select a random four-machine direction and four cross machine direction specimens measuring 100 by 200 mm (4 by 8 in.) forming the individual test sample this is followed by conditioning of the individual sample and weighing it to nearest 01 g. To preheat asphalt cement to $135 \pm 2°C$. Then to submerge the individual test specimen in the specified asphalt cement maintained at a temperature of $135 \pm 2°C$ in a mechanical convection oven. The specimen will then be submerged for 30 min two clamps may be placed on the fabric, one on each end to facilitate handling of specimen. After the required submersion, the coated asphalt cement to be removed, saturated test specimen and hang to drain (long axis vertical) in the oven at $135 \pm 2°C$. This is followed by hanging the specimen for 30 min from one end and then from the other for the same time. The asphalt cement coated, saturated test specimen is then allowed to cool for a minimum of 30 min and then trim off the excess asphalt cement. The asphalt retention is calculated as the average of the asphalt retention observed for all the specimens is as follows. $RA = (W_{sat} - W_g)/A_g$, where RA is the asphalt retention in g/m^2, W_{sat} is the weight of saturated test specimen in g, W_g is the weight of geotextile test specimen before saturation in g, and A_g is the area of geotextile specimen before test in m^2. Three samples have been tested for asphalt retention for each type of bitumen and the average value has been reported. The test results for asphalt retention are given in Table 6.4.

Table 6.4   Test results for asphalt retention of jute paving fabric

Sl. No.	Type of bitumen used for impregnation of jute paving fabric	Asphalt retention in kg/m²
1.	60/70 Bitumen	3.4
2.	PMB-40 Bitumen	3.6
3.	80/100 Bitumen	3.7

The procedure for asphalt retention test is shown in Figs. 6.1 and 6.2.

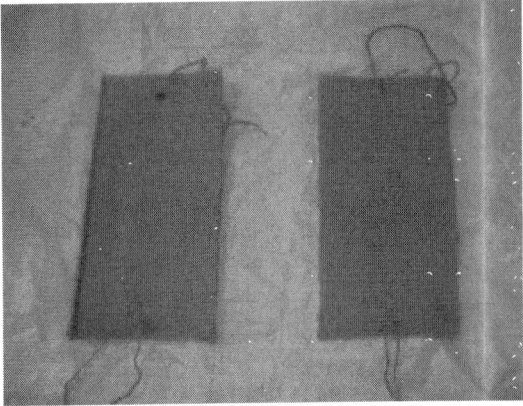

Figure 6.1   Jute samples ready for impregnation with bitumen.

Figure 6.2   Jute samples after impregnation with bitumen.

6.2.7.1 Marshall mix design method

Bruce Marshall, formerly bituminous engineer with Mississippi State Highway Department, USA formulated Marshall Method for designing bituminous mix. The test procedure has been standardized in[12] ASTM D 1559. In this method, the resistance to plastic deformations of cylindrical specimen of bituminous mixture is measured when the same is loaded at the periphery at 5 cm/min. The test procedure is used in designing and evaluating bituminous pavement mixes. The test procedure is extensively used in routine test programme for the paving jobs. There are two major features of the Marshall method of designing mixes namely density void analysis and stability-flow tests. The Marshall stability of the mix is defined as a maximum load carried by a compacted specimen at a standard test temperature at 60°C. The flow value is the deformation the Marshall Test specimen undergoes during the loading up to the maximum load, in 0.01 mm units.

6.2.7.2 Design requirements of mix as per MoRT&H specifications

As per the MoRT&H specifications for BC mix, when the specimens are compacted with 75 blows on either face, the designed BC mix should fulfil the following requirements:

Marshall stability value, kg (minimum)	900
Marshall flow value, mm	2–4
Voids in total mix, Vv %	3–6
Voids in mineral aggregates filled with bitumen, VFB, %	65–75
Loss of stability on immersion in water at 60°C	>75%

6.2.7.3 Marshall mix design and determination of OBC for the present study proportioning of aggregates

For the purpose of this study, the gradation of BC mix was selected based upon the thickness of the layer. This study was carried out for 50 mm thick layer of BC as per clause of MoRT&H specification (Fourth Revision, 2001). The individual gradation of selected component aggregates and their proportioning achieved by trial and error method is given in Table 6.5. The designed gradation along with the specified limits is shown in Fig. 6.3.

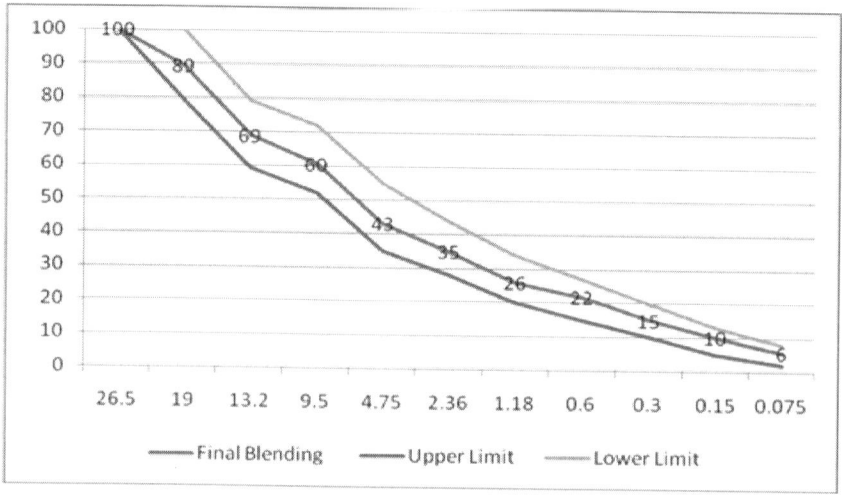

Figure 6.3   Proportioning of aggregates for BC mix design.

Table 6.5 Proportioning of aggregates for BC mix design

Sieve size	Percentage of aggregates passing through sieve size						
	Nominal size of aggregates					Blend proportion by wt. of aggregate A: B: C: D: E 31: 10: 17: 39: 3	Specified limits for 50 mm BC (MoRT &H, 2001)
	A 20 mm	B 13.2 mm	C 6 mm	D Stone dust	E Lime		
26.5	100.0	100.0	100.0	100.0	100	100	100
19	65.8	100.0	100.0	100.0	100	89	79–100
13.2	4.7	84.1	100.0	100.0	100	69	59–79
9.5	0.3	24.5	96.1	98.9	100	60	52–72
4.75	0.0	0.5	14.5	96.2	100	43	35–55
2.36	0.0	0.2	0.4	81.7	100	35	28–44
1.18	0.0	0.1	0.3	58.8	100	26	20–34
0.6	0.0	0.1	0.3	48.2	100	22	15–27
0.3	0.0	0.1	0.3	30.9	99	15	10–20
0.15	0.0	0.1	0.2	18.9	89	10	5–13
0.075	0.0	0.1	0.1	9.7	62	6	2–8

Marshall Method of the mix design as per ASTM D-1559 was carried out for determination of the optimum binder content (OBC). To determine the OBC, Marshall samples were prepared at varying percentages of 60/70 paving grade binder. Volumetric and mechanical parameters obtained for BC with 60/70 paving grade bitumen such as bulk density, Marshall stability, flow and other volumetric properties were then obtained which are given in Table 6.6. Using the above parameters, OBC was found to be 5.67% by weight of aggregates.

Table 6.6 Volumetric and mechanical parameters obtained for BC with 60/70 bitumen

Binder content, % by weight of aggregate	Bulk density, gm/cm³	Stability, kg	Flow, mm	Air voids, %	Voids filled with bitumen, VFB, %	Voids in mineral aggregates, VMA
5.0	2.382	1047	2.8	5.37	67.67	5.0
5.5	2.391	1160	3.1	4.34	74.00	5.5
6.0	2.384	1093	3.3	3.95	77.17	6.0
6.5	2.371	984	4	3.82	78.94	6.5

The values obtained at the OBC 5.67% are indicated in Table 6.7, as can be seen they do are meet MoRT&H specifications for BC mix.

Table 6.7 Marshall parameters obtained at optimum binder content with 60/70 bitumen.

Parameters	Values obtained at OBC	Specified Values as per MORT&H, 2001
Stability, kg	1160	>900
Flow, mm	3.1	2–4
Air voids, %	4.4	3–6
Voids filled with bitumen, %	73.8	65–75
Density, gm/cm³	2.390	–

6.2.8 Beam fatigue testing

The flexure fatigue test[13] is conducted to evaluate the fatigue characteristics of an HMA mixture. Fatigue cracking of pavement is considered to be more a structural problem than simply a material problem. Several external factors influence the fatigue cracking in pavements, such as poor sub-grade drainage, time of placement, and method of compaction and placement of the asphalt mix. The specimens for this test are 63.5 mm by 50 mm by 100 mm beams. The test is conducted in accordance to the procedures in AASHTO T 321-07. In this method, repeated haversine loads are applied at the third points of the specimen. The beam fatigue test can be conducted in controlled stress or controlled strain mode.

Table 6.8 Beam fatigue testing results

(A) Beam samples (with no jute)						
Strain level (microstrain)	300		400		500	
Frequency (Hz)	5	10	5	10	5	10
Number of repetitions to failure (N_f)	159,440	58,110	112,100	44,120	28,250	17,670
(B) Beam samples (with jute impregnated with 60/70 binder)						
Strain level (microstrain)	300		400		500	
Frequency (Hz)	5	10	5	10	10	5
Number of repetitions to failure (N_f)	312,320	153,200	143,640	131,250	122,590	55,040

(C) Beam samples (with jute impregnated with PMB-40 binder)						
Strain level (microstrain)	300		400		500	
Frequency (Hz)	5	10	5	10	10	5
Number of repetitions to failure (N_f)	429,510	159,900	376,900	186,430	160,120	112,090
(D) Beam samples (with jute impregnated with 80/100 binder)						
Strain level (microstrain)	300		400		500	
Frequency (Hz)	5	10	5	10	10	5
Number of repetitions to failure (N_f)	198,110	115,390	184,060	82,000	125,910	70,900

It can be seen from Table 6.8 that there is an improvement in the fatigue life of the beam where bitumen impregnated was used since they sustained more number of repetitions.

To evaluate the effect of bitumen impregnated jute in the fatigue life, a factor called 'Effectiveness Factor' (EF) has been calculated as given below:

$$\text{Effectiveness Factor (EF)} = \frac{\text{Number of repetitions to failure for reinforced beams}}{\text{Number of repetitions to failure for unreinforced beams}}$$

The effectiveness factors for the beams for different test conditions were calculated and are given in Table 6.9.

Table 6.9   Effectiveness factors for beams for different test conditions

(A) Beam samples (with jute impregnated with 60/70 binder)						
Strain level (microstrain)	300		400		500	
Frequency (Hz)	5	10	5	10	10	5
Number of repetitions to failure (N_f)	1.96	2.64	1.28	2.97	4.34	3.11
Average value of effectiveness factor (EF) = 2.72						
(B) Beam samples (with jute impregnated with PMB-40 binder)						
Strain level (microstrain)	300		400		500	
Frequency (Hz)	5	10	5	10	10	5
Number of repetitions to failure (N_f)	2.69	2.75	3.36	4.23	5.67	6.34

Average value of effectiveness factor (EF) = 4.17					
(C) Beam samples (with jute impregnated with 80/100 binder)					
Strain level (microstrain)	300		400		500	
Frequency (Hz)	5	10	5	10	10	5
Number of repetitions to failure (N_f)	1.24	1.99	1.64	1.86	4.46	4.01
Average value of effectiveness factor (EF) = 2.53					

Note: The reference beam for calculating the EF has been taken as plain beam without jute.

It can be seen from the above table that average value of EF was found to be highest for PMB-40 impregnated jute fabric. Also, PMB-40 gives higher values of EF for all the strain levels and frequency loadings. So, it can be concluded the PMB-40 is the most effective binder for increasing the fatigue life and will mitigate the propagation of reflective cracking. However, field performance evaluation is a must for evaluating the actual behaviour under ambient climatic conditions.

6.2.9 Wheel tracking test

Wheel tracking is used to assess the resistance to rutting of asphaltic materials under conditions which simulate the effect of traffic. A loaded wheel tracks a sample under specified conditions of load, speed and temperature while the development of the rut profile is monitored continuously during the test. The wheel tracking test consists of a loaded wheel assembly and a confined mould in which a 305 × 305 × 50 mm specimen of asphalt mix is rigidly restrained on its four sides. The test specimen is mounted on a table which is reciprocated a distance of 230 mm on linear bearings at the specified speed of 42 passes/min along the length of the slab. A loaded rubber tyred wheel runs on top of the specimen and the resultant rut is monitored as the test proceeds using a calibrated displacement transducer. The temperature during the test is maintained by an insulated closed chamber maintained at a constant test temperature of 50 ± 1°C. The specimens are subjected to 20,000 cycles. Two specimens were tested for each mix and average data on rut depth was found out. The rut depth was recorded at mid-point of the specimen length. The slabs for this test were prepared by filling the mould with the bituminous mix and applying static load through UTM till the depth of 50 mm is achieved. Two different types of slabs were prepared for this test one is the slab with control mix (in which no jute was used) and the other is the slab in which jute impregnated with PMB-40 was laid in the bottom one-third height of the sample. The results for the wheel tracking test are plotted in Fig. 6.4.

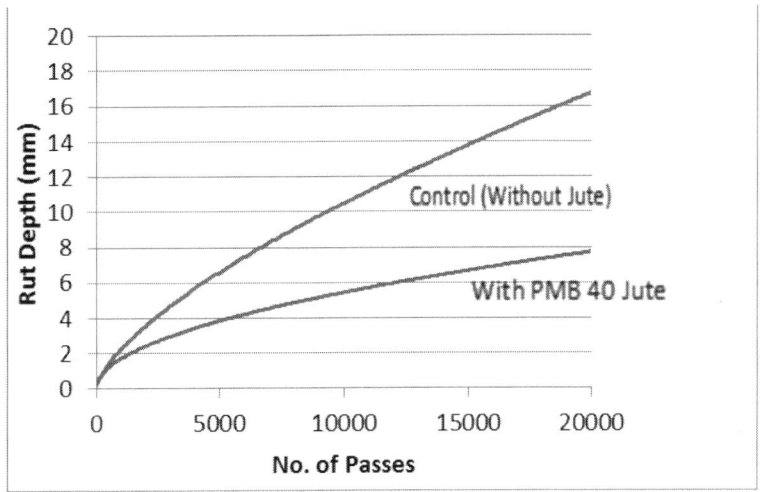

Figure 6.4   Number of passes vs. rut depth (in wheel tracking test).

6.3   Bituminization of grey jute paving fabric (GJPF)

After thorough laboratory testing of the compatibility of different types and grades of bitumen with grey jute fabric, the CRRI scientists observed and recommended that jute was found to be effective in increasing the fatigue life of bituminous mixes. Jute impregnated with 60/70 bitumen was found to have an average EF of 2.72, i.e. it increases the fatigue life by 172% compared to samples where no jute was used. Jute impregnated with PMB-40 bitumen was found to have an average EF of 4.17, i.e. it increases the fatigue life by 317% compared to samples where no jute was used. Finally, jute impregnated with 80/100 bitumen was found to have an average EF of 2.53, i.e. it increases the fatigue life by 153% compared to samples where no jute was used. A higher value of EF indicates higher potential of the developed fabric to be used for field trial as per the objectives of the project. Based on the laboratory testing and analysis of the results obtained, the CRRI scientists recommended that jute impregnated with PMB-40 bitumen was found to have the highest fatigue life and therefore, it is recommended to be used for the purpose of field trials. Proper impregnation of the jute fabric as per ASTM 6140 needed to be ensured. It was recommended to be confirmed that the full thickness of the jute fabric is impregnated with the selected bitumen along with the other necessary chemicals. However, any excess bitumen on the surface of jute fabric should be removed immediately. On the basis of the recommendations made by the CRRI scientists, applying PMB-40 bitumen on the grey jute paving fabric (GJPF), production of bituminized jute paving fabric (BJPF) has been carried out in a Bitumen Treatment Plant, Kolkata. The process flow chart along with Figs. 6.5–6.11 and the technical data have been furnished in Table 6.10, respectively.

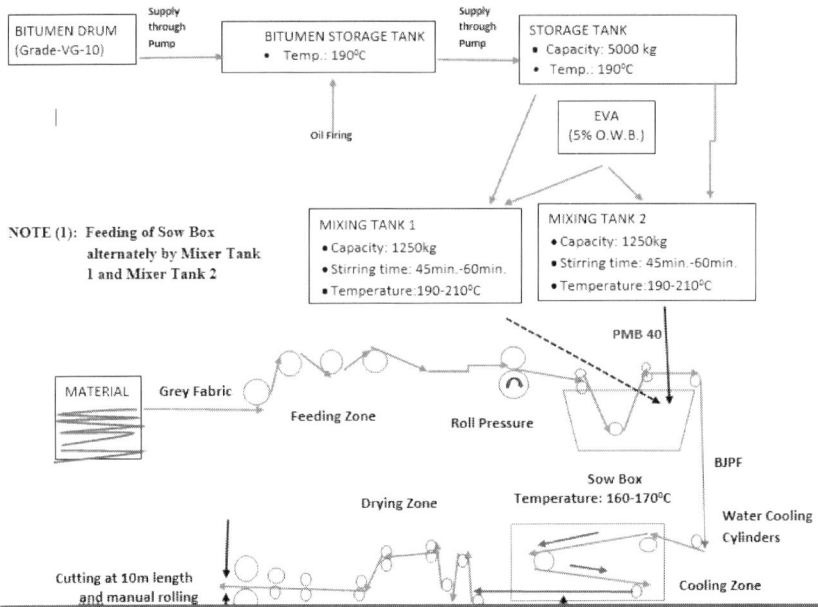

Figure 6.5   Process flow chart showing bituminization of grey jute paving fabric (GJPF).

Table 6.10   Technical data during bituminization of the grey jute 5-layered combined paving fabric in the sow box

Sl. No.	Property parameters and particulars	Values
1	Sow box temperature	160°C–170°C
2	Bitumen take-up % of the fabric before being squeezed	320%
3	Bitumen take-up % of the fabric after passing through the nip of squeeze roller-mangle roller assembly	240%
4	Bitumen content	0.24 g/cm²
5	Bitumen squeeze out	80%
6	Nip roller pressure	3.0 kg/cm²
7	Distance between the squeeze roller and mangle roller	5 mm

The specifications of the final BJPF that has developed by bituminizing grey jute fabric sample have been furnished in Table 6.11.

Figure 6.6 Stacking of GJPF before bituminization.

Figure 6.7 Feeding of GJPF into the sow box before bituminization.

Figure 6.8 Bitumen PMB-40 impregnation into GJPF during bituminization.

Figure 6.9 Dipping of GJPF into the sow box during bituminization.

Figure 6.10 Bituminization of GJPF going on.

Figure 6.11 Delivery Zone-developed BJPF getting enrolled.

Table 6.11   Specifications of the bituminized jute paving fabric (BJPF)

Sl. No.	Parameters	Values
1.	Fabric weight expressed in gsm	3500.00
2.	Thickness expressed in mm	8.00
3.	Wide-width tensile strength expressed in kN/m (machine direction × cross-machine direction)	38.50 × 40.00
4.	Elongation at break expressed in percentage (machine direction × cross-machine direction)	11.0 × 9.0
5.	Bursting strength expressed in kgf/cm²	43.00

References

1. Specifications for Road and Bridges Works—Fourth Revision, Ministry of Road Transport and Highways, MoRT&H, Indian Roads Congress, New Delhi, 2001.

2. Fatigue of Compacted Bituminous Aggregate Mixtures—Special Publication 508. In: Proc. Symposium presented at the Seventy-fourth Annual Meeting, American Society for Testing and Materials, ASTM, Atlantic City, NJ, 1971.

3. The Asphalt Handbook, Manual Series, MS-4, seventh edition, Asphalt Institute, USA, 2007.

4. Marienfield, M.L. and Baker, T.L., Paving Fabric Interlayer System as a Pavement Moisture Barrier, Transportation Research Circular Number E-C006, TRB, National Research Council, Washington, DC, 1999.

5. Methods of Test for Aggregates for Concrete–Particle Size and Shape, Bureau of Indian Standards, IS: 2386(Part 1)-1963 (Reaffirmed 2002), New Delhi, 2002.

6. Method of Test for Determination of Stripping Value of Road Aggregates, Bureau of Indian Standards, IS: 6241-1971 (Reaffirmed 2003), New Delhi, 1971.

7. Don, L.L., Physical Geology, sixth edition, Prentice-Hall, Englewood Cliffs, NJ, pp. 5–10, 1982.

8. Methods for Testing Tar & Bituminous Materials: Determination of Ductility, Bureau of Indian Standards, IS: 1208, New Delhi, 1978.

9. Methods for Testing Tar & Bituminous Materials: Determination of Softening Point, Bureau of Indian Standards, IS: 1205, New Delhi, 1978.

10. Methods for Testing Tar & Bituminous Materials: Determination of Specific Gravity Penetration, Bureau of Indian Standards, IS: 1202, New Delhi, 1978.

11. Methods for Testing Tar & Bituminous Materials: Determination of Penetration, Bureau of Indian Standards, IS: 1203, New Delhi, 1978.

12. Standard Test Method for Resistance to Plastic Flow of Bituminous Mixtures using Marshall Apparatus, American Society for Testing and Materials, ASTM D 1559, Atlantic City, NJ, 2010.

13. Adhikari, S. and Zhanping, Y., Fatigue evaluation of asphalt pavement using beam fatigue apparatus, Technology Transfer Journal, 10(3), 1–3, 2010.

Pilot and bulk field trials of bituminized jute paving fabric (BJPF)

7.1 Introduction

In general, in field trials a newly developed product is tested by users in a real life setting as opposed to testing under artificial laboratory conditions. Both the product and the field trial setting are designed to be as close as possible to actual usage. In this context, field trial, both in pilot and bulk forms, involves installation of developed bituminized jute paving fabric (BJPF) in urban road and state highway of high traffic volume followed by its monitoring over a period of time. The result of such investigation contains valuable information for both the practicing civil engineers as well as textile engineers regarding the potential for improving the usability of the product. It is worthy to mention here that the use of such field trials is fairly useful for the testing of new products prior to their commercial launch.

7.2 Pilot field trial

Pilot field trial of the developed BJPF has been carried out at the premises of the Department of Jute and Fibre Technology, Institute of Jute Technology, University of Calcutta. The laying of the bitumen impregnated jute fabric has been done in accordance with the requirements of IRC: SP: 59-2002. Monitoring of the field trial and performance evaluation of the BJPF are going on which will be under a constant observation for at least two years. Close monitoring of the field trial has been conducted after every 15 days and this continues for the next two years to assess the performance of the road under field trial. In the course of physical observation during monitoring of the road section under traffic simulated condition there were no signs of cracks and pot holes appearing on the surface of the road even after the completion of ten months of traffic simulation programme. No water logging is observed after rainfall. Standardization and optimization of the product will be made and disseminated to the manufacturers and end users for commercial application after the completion of monitoring and performance evaluation of the product. Figures 7.1–7.8 of this pilot field trial are shown below.

Figure 7.1   Cleaning of the site in the departmental campus.

Figure 7.2   Preparation of bituminous mixture.

Figure 7.3   Laying of tack coat.

Figure 7.4   Installation of bituminized jute paving fabric (BJPF).

Figure 7.5   Laying of mastic over BJPF.

Figure 7.6   Placing of anti-skid stone chips.

Figure 7.7 Thermosealing of the joints.

Figure 7.8 Newly masticated road section

7.3   Commercial field trial – site 1

Based on the pilot field trial performance, a full-scale commercial field trial has been carried out over a stretch of lane of area −1000 m² approximately at Uday Shankar Sarani, Golf Green, Kolkata 700095 to evaluate both the functional and structural contribution of the developed BJPF reinforcement to the pavement system in a full volume traffic road. The detailed location of the site is furnished below:

Site: Uday Shankar Sarani, Golf Green, Kolkata 700 095

Location: 22.4931499′N and 88.3633918′E

Sub-locality: Golf Green

Country: India

State: West Bengal

District: Kolkata

Locality: Kolkata

The developed BJPF has been carried to the site and laid on the ear marked road as per the following sequence.

1. Ear-marking of the stretch of road
2. Cleaning the surface
3. Cooking of the ingredients in the vat to prepare the mastic
4. Application of the tack coat on the cleaned surface of the road before laying the fabric
5. Repairing of the patches on the road surface

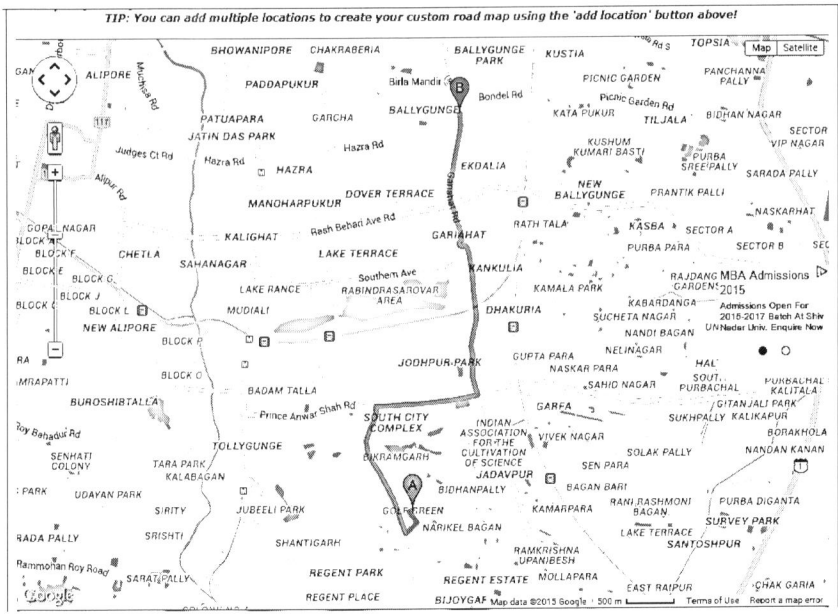

Figure 7.9 Map showing the location of the field trial site-1, Uday Shankar Sarani, Golf Green, Kolkata 700 095, West Bengal, India.

6. Laying of the BJPF

7. Application of the mastic (thickness 20 mm) on the fabric

8. Placing of anti-skid stone chips in definite intervals on the newly prepared road

9. Thermosealing of the portions of mastic along with BJPF

Table 7.1 Ingredients and their proportion required in every 1260 kg preparation of mastic in the vat (as per the standard recipe prescribed by Public Works Department, Government of West Bengal)

Sl. No.	Ingredients	Quantity (kg)
1.	Bitumen (90:15) (VG-40)	130
2.	Sand	200
3.	Stone chips (1/2 in. down)	500
4.	Stone chip dusts	140
5.	Lime dust	290

Close monitoring of the ear-marked road is going on at regular intervals and the result obtained so far is fairly satisfactory. Even after withstanding two monsoonal seasons, except some minor cracks, no major distress in the form of potholes or major cracks or any other irregularity has been observed so far. Traffic is moving smoothly over the experimented road portion. Figures 7.10–7.19 of this commercial field trial are shown below.

Figure 7.10 Melting of bitumen and cooking of ingredients in the vat.

Figure 7.11 Cleaning of the road surface.

Figure 7.12 Application of tack coat.

Figure 7.13 Laying of bituminized jute paving fabric.

Figure 7.14 Transfer of mastic over laid fabric.

Figure 7.15 Placing of anti-skid stones (view-1).

Figure 7.16 Placing of anti-skid stones (view-2).

Figure   7.17 Analysing   road surface roughness by Merlin Tester.

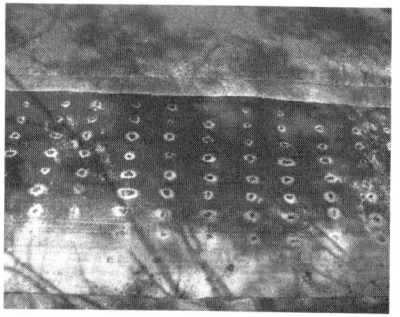

Figure 7.18 View of the newly prepared road surface.

Figure 7.19 Vehicles running on the newly masticated road.

7.4   Commercial field trial – site 2

Site: Contai-Beldah Carriageway, State Highway-5, West Bengal, India

Location: 21.781587′N, 87.736825′E

Sub-Location: Paschim Medinipur, West Bengal

Country: India

State: West Bengal

District: West Medinipur

Locality: Medinipur

Realizing the potentiality of the developed BJPF material along with all its field trial reports and in order to establish the techno-economic feasibility of the developed product through more number of such potential field applications, National Jute Board (NJB), Ministry of Textiles, Government of India has approached the Chief Engineer, PWD, West Bengal with proposals to take some more commercial field trials of BJPF in high volume traffic roads like State Highways. Superintending Engineer, PWD, South Western Circle, responded to this proposal of NJB by sanctioning a stretch of road from 36 kmp to 37 kmp at State Highway-5, Contai-Beldah Carriageway, Paschim Medinipur for strengthening the carriageway with the developed BJPF. Accordingly, 3600 m^2 of BJPF as per the recommendations and specifications furnished by Central Road Research Institute (CRRI), New Delhi have been laid down in the ear-marked stretch. The trial commenced from November 17, 2013 under the supervision of Department of Jute and Fibre Technology, University of Calcutta, India and National Jute Board, MoT, GoI along with the PWD officials. The assigned supervisor of the work as well as the workers engaged in the field trial activity have been guided and trained by the department faculties and NJB personnel, specifically regarding the tack coat application followed by laying of the BJPF and application of mastic asphalt on the laid BJPF. After the completion of the experiment, the road traffic started its movement regularly. So far the first-hand information received from the monitoring desk as well as from the contractor's end, that the regular traffic are passing through the experimental set-up of the road without facing any disturbance and the stretch of road is performing satisfactorily. Figures 7.20–7.28 of this commercial field trial are shown below.

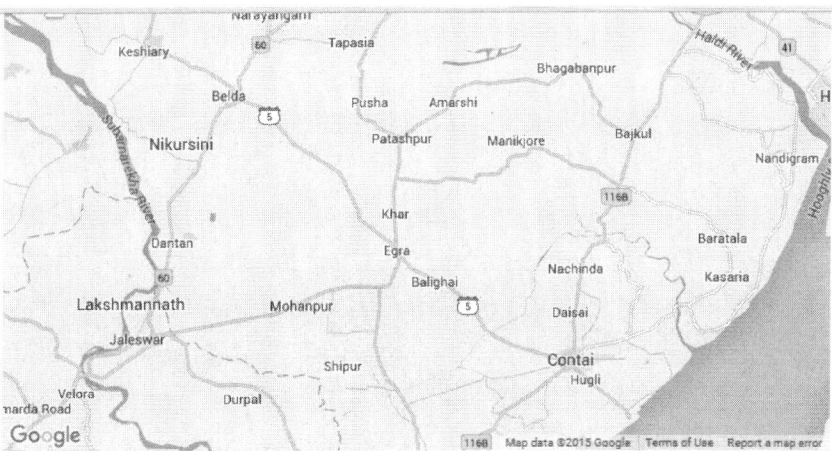

Figure 7.20   Map showing the location of the field trial site-2, Contai-Beldah Carriageway, State Highway-5, West Bengal, India.

Figure 7.21 State Highway 5 before fabric application.

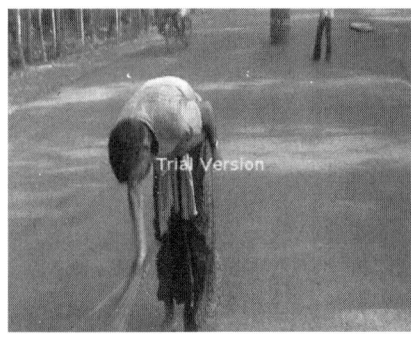

Figure 7.22 Cleaning of the road surface.

Figure 7.23 Accumulation of ingredients before cooking.

Figure 7.24 Cooking of ingredients in the vat.

Figure 7.25 Application of tack coat before laying of BJPF.

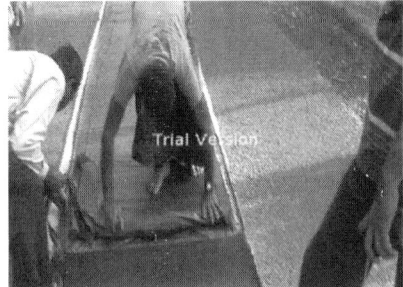

Figure 7.26 The laying of BJPF on the cleaned surface.

Figure 7.27 Laying of mastic over the paving fabric surface.

Figure 7.28 Placing of anti-skid stone chips over masticated road surface.

Reinforcement, 8, 9, 35, 36, 110–112
Reinforcing, 25, 38, 75
Retting, 23, 30
Rotating thin film oven test (RT-FOT), 91, 92
Rotational Viscometer, 86, 87
Rutting, 76, 88, 102, 109, 120

S

Sandwich fabrics, 49, 50
Satin weave, 16, 17
Separation, 6, 7, 33, 36, 109
Shape test, 127
Short Staple System, 31
Short-term yarn irregularity, 32
Softening point temperature, 87
Soil conditions, 1
Soil saver, 25, 34, 36, 37, 41
Soil technology, 1
Specific gravity test, 96
Spun bonded fabrics, 19
Spun bonding, 4
Spunlaid fabrics, 19
Stabilizing, 10, 38
Staple fibres, 4, 19, 23
Stiffness, 16, 36, 98, 117
Strain Absorbing Membrane Interlayer (SAMI), 109, 110
Stripping test, 127
Strong acids, 26
Styrene Isoprene Styrene (SIS), 81, 103
Synthetic geotextile, 3, 33, 36

T

Tenacity, 25, 36, 50
Tensile strength, 13, 32, 33, 36, 38, 52, 53, 62, 67
Thermal cracking, 98, 102
Three-dimensional (3D) mats, 4, 36
Torsional rigidity, 25, 32
Tossa jute, 23, 25, 28
Triaxial weave, 16, 17
Twill weave, 16, 17

U

Unevenness, 31

V

Viscosity grade bitumen, 82
Viscosity test, 83, 85, 128

W

Warp yarns, 4, 16, 37
Water absorption test, 125
Water permeability, 7, 68–70
Weft yarns, 4, 16
Wheel tracking test, 134, 135
Winding, 31, 32
Woollenization, 26
Woven geotextiles, 4, 15, 16